Contributions to Statistics

For further volumes:
http://www.springer.com/series/2912

Silvestro Montrone • Paola Perchinunno

Editors

Statistical Methods for Spatial Planning and Monitoring

Springer

Physica-Verlag

A Springer Company

Editors
Silvestro Montrone
Paola Perchinunno
Dipartimento di Studi Aziendali e Giusprivatistici
University of Bari
Bari
Italy

ISSN 1431-1968
ISBN 978-88-470-5607-7   ISBN 978-88-470-2751-0 (eBook)
DOI 10.1007/978-88-470-2751-0
Springer Milan Heidelberg New York Dordrecht London

© Springer-Verlag Italia 2013
Softcover reprint of the hardcover 1st edition 2013
This work is subject to copyright. All rights are reserved by the Publisher, whether the whole or part of the material is concerned, specifically the rights of translation, reprinting, reuse of illustrations, recitation, broadcasting, reproduction on microfilms or in any other physical way, and transmission or information storage and retrieval, electronic adaptation, computer software, or by similar or dissimilar methodology now known or hereafter developed. Exempted from this legal reservation are brief excerpts in connection with reviews or scholarly analysis or material supplied specifically for the purpose of being entered and executed on a computer system, for exclusive use by the purchaser of the work. Duplication of this publication or parts thereof is permitted only under the provisions of the Copyright Law of the Publisher's location, in its current version, and permission for use must always be obtained from Springer. Permissions for use may be obtained through RightsLink at the Copyright Clearance Center. Violations are liable to prosecution under the respective Copyright Law.
The use of general descriptive names, registered names, trademarks, service marks, etc. in this publication does not imply, even in the absence of a specific statement, that such names are exempt from the relevant protective laws and regulations and therefore free for general use.
While the advice and information in this book are believed to be true and accurate at the date of publication, neither the authors nor the editors nor the publisher can accept any legal responsibility for any errors or omissions that may be made. The publisher makes no warranty, express or implied, with respect to the material contained herein.

Printed on acid-free paper

Physica is a brand of Springer
Springer is part of Springer Science+Business Media (www.springer.com)

Preface

In an ever more rapidly evolving modern society with less and less standardized models of behavior, the government of the territory and of the dynamics that determine its evolution appears essential for the pursuit of objectives of sustainable development.

The uncertainty that characterizes the socioeconomic and territorial dynamics of contemporary society requires the elaboration of forecasting methods and instruments that are able to manage and govern the complexity (I. Prigogine 1986 and L. von Bertalanffy 1971).

In the absence of correct planning, the rapid growth of cities can provoke the phenomena of "social exclusion." The ability to govern the transformation is linked to the country's economic, social, technological, cultural, and political prospects of growth (L. Balbo 1992). It is therefore necessary to adopt a theoretical approach to the problem that takes account of how the different scenarios change in the course of time.

The measurement of territorial phenomena is therefore fundamental to the cognitive model on which we base the choice of territorial policies and urban plans. The construction of statistical indicators permits the integration of the different social, economic, environmental, and urban characteristics of a territory and highlights their relationships or dependence at a territorial level.

This book aims to investigate methods and techniques for spatial statistical analysis suitable to model spatial information in support of decision systems.

Over the last few years there has been a considerable interest in these tools and in the role they can play in spatial planning and environmental modeling. One of the earliest and most famous definitions of spatial planning was "a geographical expression to the economic, social, cultural, and ecological policies of society" (European Conference of Ministers Responsible for Regional/Spatial Planning). Borrowing from this point of view, this text shows how an interdisciplinary approach is an effective way to a harmonious integration of national policies with regional and local analysis.

The book covers a wide range of spatial models and techniques: spatial data mining, point processes analysis, nearest neighbor statistics and cluster detection, fuzzy regression model, and local indicators of spatial association. All of these tools provide the policy-maker with a valuable support to policy development.

Chapter 1 covers basic and more advanced aspects of Bayesian hierarchical modeling for disease mapping. It also describes the methods for the analysis of whether the spatial distribution of the disease risk closely follows that of the underlying population at risk or whether there are some nonrandom local patterns (disease clusters) which may suggest a further explanation for disease etiology. The case study examined concerns the spatial distribution of liver cancer mortality in Apulia.

In Chap. 2 starting from Diamond (1988), defined in the space of triangular fuzzy numbers, in the context of a simple linear regression model, a multivariate generalization is proposed, through the indication of a "stepwise" method for the selection of variables. As an application case of the proposed measure of dissimilarity, homogeneous groups of Italian universities are identified, according to graduates' opinions on many aspects concerning internship activities.

Chapter 3 enlarges the use of variogram-based geostatistical techniques to analyze time series. In order to underline the role of the variogram for modeling and prediction purposes, several theoretical aspects, such as interpolation of missing values, temporal prediction, nonparametric estimation, and their computational problems, are faced through an extensive case study regarding an environmental time series.

In Chap. 4 spatiotemporal geostatistical analysis is combined with the use of a Geographic Information System (GIS): the integration of geostatistical tools and GIS enables the identification and visualization of alternative scenarios regarding a phenomenon under study and supports environmental risk management. The case study is based on environmental data measured at different monitoring stations in the southern part of the Apulia Region (South of Italy).

Chapter 5 compares two different clustering methods: the first based on the technique of SaTScan and the other based on the use of Seg-DBSCAN, a modified version of DBSCAN. The main objective is to identify territorial areas characterized by situations of deprivation or strong social exclusion through a fuzzy approach. Grouping methods for territorial units are employed for areas with high intensity of the phenomenon by using clustering methods that permit the aggregation of spatial units that are both contiguous and homogeneous with respect to the phenomenon under study.

Chapter 6 focuses its attention on the topic of "shrinkage," spatial phenomenon defined by data and information based on space dimension relying on spatial information. The wide use of geo-information is a useful aid to extend common statistical analyses by integrating data collected at different levels, comparing data at a municipal level to data referring at census area level. The paper includes an empirical section describing the case of the de-industrialized city of Taranto, measuring the major indicators of shrinkage.

Chapter 7 describes a procedure for investigating the coherence of the relationship between a "wide" concept of spatial distance from some reference point and the geographical variation of real estate value. In the case study on the main districts of New York, it is possible to investigate the effect of racial steering on ethnic dissemination and real estate variation.

Finally, in the case study, a variety of themes covered, both from theoretical and methodological points of view, find a real useful application to immediately understand the models illustrated.

We would like to thank all the authors for their contributions.

Bari, Italy   Silvestro Montrone and Paola Perchinunno

Contents

Chapter 1
Geographical Disparities in Mortality Rates: Spatial Data Mining and Bayesian Hierarchical Modeling

Massimo Bilancia, Giusi Graziano, and Giacomo Demarinis

Abstract Achieving health equity has been identified as a major international challenge since the 1978 declaration of Alma Ata. Disease risk maps provide important clues concerning many aspects of health equity, such as etiology risk factors involved by occupational and environmental exposures, as well as gender-related and socioeconomic inequalities. This explains why epidemiological disease investigation should always include an assessment of the spatial variation of disease risk, with the objective of producing a representation of important spatial effects while removing any noise. Bearing in mind this goal, this review covers basic and more advanced aspects of Bayesian models for disease mapping, and methods to analyze whether the spatial distribution of the disease risk closely follows that of underlying population at risk, or there exist some nonrandom local patterns (disease clusters) which may suggest a further explanation for disease etiology. We provide a practical illustration by analyzing the spatial distribution of liver cancer mortality in Apulia, Italy, during the 2000–2005 quinquennial. (Massimo Bilancia wrote Sects. 1.1.2, 1.1.4, 1.1.6, 1.2.1, 1.2.3, 1.2.5. Giusi Graziano wrote Sects. 1.1.1, 1.1.3, 1.1.5, 1.2.2, 1.2.4, 1.2.6. Giacomo Demarinis wrote the software for data analysis. Section 1.3 was written jointly. The three authors read and approved the final manuscript. We wish to thank Maria Rosa Debellis, Department of Neuroscience and Sense Organs, University of Bari, Italy, and Claudia Monte PhD, Department of Physics, University of Bari, Italy, for their valuable support.)

M. Bilancia (✉) • G. Demarinis
Dipartimento di Scienze Statistiche "Carlo Cecchi", University of Bari "Aldo Moro",
Largo Abbazia di Santa Scolastica 53, 70124 Bari, Italy
e-mail: mabil@dss.uniba.it; demarinis.giacomo@gmail.com

G. Graziano
Laboratory of Lipid Metabolism and Cancer, Consorzio Mario Negri Sud, Via Nazionale 8/A,
66030 Santa Maria Imbaro (CH), Italy
e-mail: graziano@negrisud.it

S. Montrone and P. Perchinunno (eds.), *Statistical Methods for Spatial Planning and Monitoring*, Contributions to Statistics, DOI 10.1007/978-88-470-2751-0_1, © Springer-Verlag Italia 2013

Keywords Bayesian statistics • Besag–York–Mollié model • Disease cluster detection • Disease mapping • Spatial scan statistic

1.1 Statistical Thinking in Spatial Epidemiology

1.1.1 Health Equity in a Spatial Setting

Health equity concerns the study of differences in health and health care across different populations: achieving health equity has been identified as a major international challenge since the 1978 declaration of Alma Ata. Of course, inequalities need to be quantified before they can be addressed, and since the mid-1800s maps have been commonly used to provide a visual representation of disease outcomes, geographical variability in mortality and incidence, or treatment and survival. For example, atlases of cancer incidence and mortality provide important clues concerning many aspects of health equity, such as etiological risk factors related to occupational and environmental exposures, or gender-related and socioeconomic inequalities, as well as barriers to enter in health-care delivery systems or the quality of primary cancer control factors which different ethnic and racial groups receive [11].

This explains why epidemiological disease investigation should include, whenever possible, an assessment of the spatial variation of disease risk, with the objective of producing a visual representation of important spatial effects while removing any disturbing noise. For aggregate province or district-level data, disease mapping has a long history in epidemiology: after briefly reviewing Gaussian Markov random fields (GMRFs), which are frequently used in statistics and are a basic building block in spatial statistics, we examine the foundations of Bayesian modeling in disease mapping in Sects. 1.1.3 and 1.1.4. Spatially aggregated data will be the main focus of this paper, whereas no attention will be paid to point patterns, for which we send back to some excellent reviews (see, for example, [29]).

The rest of the paper is organized as follows: in Sect. 1.1.5 we consider a related but equally important problem for public health, that is whether the presence of high incidence/mortality rates within a group of neighboring areas means a higher risk of suffering from the disease under study (the so-called "local cluster detection" problem). In Sect. 1.1.6, we bridge the gap between Bayesian disease mapping and local cluster detection methods, with the help of a newly introduced suitable modification of the Besag–York–Mollié model with ecological covariates to scan spatial disease rates. The second part of the paper provides a practical illustration of the methodologies discussed in the first part: in particular, the spatial distribution of liver cancer mortality in Apulia, Italy, during the 2000–2005 quinquennial is analyzed. Finally, current limitations and future prospects are briefly discussed in the conclusions of the paper.

1.1.2 Gaussian Markov Random Fields

GMRFs are frequently used in a variety of fields, in particular in disease mapping for modeling spatial dependence [33] this is due to the fact that risk levels of areas that are close to each other will often tend to be positively correlated as they share a number of spatially varying characteristics. In this context, we suppose to have N overlapping areas with the vector $\psi = (\psi_1, ..., \psi_N)$ representing the effects describing the spatially structured variability in the logarithm of area-specific relative risks. Why log-relative risks should be modeled, rather than the risks themselves, will be apparent in the next section.

In this section, we are facing with the problem of modeling this N-dimensional random variable considering the dependence between ψ_i and ψ_j in a suitable way, where i and j are two different areas. There are two common approaches to specify a distribution for ψ: the joint modeling and the conditional one. In the first case, assume that ψ is distributed according to the following multivariate Gaussian random field

$$\psi \sim N\left(0, \ \sigma_\psi^2 \Sigma\right) \tag{1.1}$$

where σ_ψ^2 is the common variance of the ψ_i and Σ is a $N \times N$ positive definite correlation matrix whose off-diagonal terms describe the dependence between ψ_i and ψ_j. The matrix Σ can assume different forms [35] but its elements Σ_{ij} are usually expressed as a function of the distance d_{ij} between the centroids of area i and j

$$\Sigma_{ij} = f(d_{ij}, \lambda) = \exp\left\{-\left(\frac{d_{ij}}{\lambda_1}\right)^{\lambda_2}\right\} \tag{1.2}$$

where $\lambda_1 > 0$ controls the extent of the spatial dependence and $\lambda_2 \in [0, 2)$ is a smoothing parameter. The joint modeling is often computationally expensive and requires the specification of the elements of the covariance matrix Σ. As reviewed in [4], starting from the standard properties of the multivariate Normal, the joint specification (1.1) is equivalent to the following set of conditional distributions

$$\psi_i | \psi_j \sim N\left(\sum_{j=1}^{N} W_{ij} \psi_j, \ \sigma_\psi^2 D_{ii}\right) \tag{1.3}$$

with $W_{ii} = 0$, $W_{ij} = -Q_{ij}/Q_{ii}$ and $D_{ii} = Q_{ii}^{-1}$, where Q_{ij} is the (i, j)-element of the matrix $Q = \Sigma^{-1}$. The two approaches are equivalent to each other, in the sense that the conditional distributions (1.3) require the specification of the matrix W of weights W_{ij} and the diagonal matrix D whose diagonal elements are D_{ii}, and lead to the joint specification

$$\psi \sim N\left(0, \ \sigma_\psi^2 (I - W)^{-1} D\right) \tag{1.4}$$

provided that the compatibility conditions $W_{ij}D_{jj} = W_{ji}D_{ii}$ are satisfied to ensure the symmetry of $Q = D^{-1}(I - W)$. A particularly useful GMRF, also known as *conditional autoregression* (CAR) model, is obtained by taking $W_{ij} \neq 0$ in (1.3) if areas i and j are neighbors (abbreviated as $i \sim j$) and $W_{ij} = 0$ otherwise (including the case that $W_{ii} = 0$). Of course, the simplest definition is $W_{ij} = 1$ if $i \sim j$ and $W_{ij} = 0$ otherwise [3, 44]. In this case, denoting with m_i the number of areas neighboring to area i, it is quite simple to show that $D_{ii} = m_i^{-1}$ and $W_{ij} = m_i^{-1}$ and that (1.3) reduce to

$$\psi_i | \psi_j \sim N\left(\frac{1}{m_i}\sum_{j \in \partial_i} \psi_j, \ \frac{\sigma_\psi^2}{m_i}\right) \tag{1.5}$$

where ∂_i is the set of neighbors of area i, the conditional mean of ψ_i is the average of the neighboring ψ_i's and the conditional variance is inversely proportional to the number of neighbors m_i. The model (1.5) has been proposed in [2], and it is commonly referred to as *intrinsic conditional autoregression* (ICAR) model. It is a limiting form of the CAR model defined above because the covariance matrix Q results to be not positive definite. For this reason σ_ψ^2 is only interpretable conditionally and no longer as a marginal variance (because the joint specification no longer exists). The ICAR model, thanks to Markov Chain Monte Carlo (MCMC) methods, is very popular in the analyses of areal data. Strategies to address the theoretical and computational difficulties of CAR models have been proposed [1]; however, the research of other methods is object of ongoing studies.

1.1.3 Bayesian Thinking in Spatial Disease Mapping

Let Y_{ij} be the number of disease counts or deaths for a specific cause within area i ($i = 1, 2, \ldots, N$), classified according to stratum j ($j = 1, 2, \ldots, J$, e.g., sex and age classes). For rare disease we assume that, independently in each area and stratum and conditionally on stratum-specific rates p_{ij}, counts are distributed as

$$Y_{ij} | p_{ij} \overset{ind}{\sim} \text{Poisson}(R_{ij}p_{ij}) \tag{1.6}$$

where p_{ij} is the stratum-specific mortality rate within the ith area, and R_{ij} is the number of person-years at risk. It is customary to reduce the dimension of the parameter space by assuming that the following proportionality relation holds for the $N \times J$ probabilities p_{ij}

$$p_{ij} = q_j \times \theta_i \tag{1.7}$$

where q_j, $j = 1, 2, \ldots, J$, is a set of stratum-specific reference rates and $\theta_i = p_{ij}/q_j$ is interpreted as a relative risk (rate ratio) associated with area i. The proportionality assumption (1.7) is discussed in detail by Dabney and Wakefiled [12]: an examination of the literature reveals that checking whether (1.7) is valid seems to be rarely reported. Exploiting the reproductive property of the Poisson distribution, we may collapse over strata to obtain the following saturated Poisson model for area-specific counts $Y_i = \Sigma_j Y_{ij}$

$$Y_i | \theta_i \overset{ind}{\sim} \text{Poisson}(E_i \theta_i) \tag{1.8}$$

where the expected value is obtained as $\Sigma_j R_{ij} p_{ij} = \Sigma_j R_{ij} q_j \times \theta_i = E_i \theta_i$ and $E_i = \Sigma_j R_{ij} q_j$, that is E_i is the *expected* number of cases in area i, which accounts for effects attributable to differences in the confounder-specific populations, and seeks to answer the question of what would be the number of cases expected in the study population if people contract the disease at the same rate as people in the standard population (*indirect standardization*). Waller and Gotway [45] and Ocaña-Riola [25] warn against the indiscriminate application of the indirect standardization and point out some possible inaccuracies arising, for example, from the misuse of total population as the denominator for incidence or mortality rates from a specific cause.

When the reference population is the same as the population under study we set the stratum-specific reference rates as

$$q_j = \frac{\Sigma_i Y_{ij}}{\Sigma_i R_{ij}}, \quad j = 1, 2, \dots, J. \tag{1.9}$$

This approach centers the data with respect to the current map, and it is a simple exercise to verify that $\Sigma_i Y_i = \Sigma_i E_i$ under indirect internal standardization. The areas where there is an excess of risk are those in which the number of observed cases exceeds the expected one.

Maximum likelihood estimates of area-specific relative risk in model (1.8) are given by $\hat{\theta}_i = \text{SMR}_i = Y_i / E_i, \quad i = 1, 2, \dots, N$, where SMR_i is commonly referred to as the *standardized mortality ratio*. Exact area-specific p-values associated with the null hypothesis of no increased risk H_0: $\theta_i = 1$ against H_i: $\theta_i > 1$ are given by

$$\rho_i = \Pr\{x \geq Y_i | E_i\} = 1 - \sum_{x=0}^{Y_i - 1} \frac{\exp(E_i) E_i^x}{x!}. \tag{1.10}$$

All of these values or their complements $1 - \rho_i (i = 1, 2, \dots, N)$, as well as threshold probabilities from the right-continuous distribution function $1 - \tilde{\rho}_i = \Pr\{x \leq Y_i | E_i\}$, may be classified to draw a probability map attributing to each area a gray or color level that denotes class membership.

It is worth noting that probability maps may be not very informative, as p-values alone do not give any information about the level of risk. What is worst, outcomes in spatial units are often not independent from each other: risk levels of areas that are close to each other will tend to be positively correlated as they share a number of spatially varying characteristics: ignoring the over dispersion caused by spatial autocorrelation will lead to incorrect inferences. In other words, the marginal variance of counts Y_i will be systematically larger than the variance resulting from the independent-component Poisson model (1.8), in which $Var(Y_i) = E_i\theta_i$ equals $E(Y_i)$: extreme p-values might be due more to the lack of fit than to a real risk excess.

If we admit that variation of the observed number of events in a given area undergoes some degree of extra-Poisson variation, we can summarize this by a prior distribution $\pi(\theta|\eta)$, containing information about the variability in the relative risks across the map, which depends on a vector of hyperparameters η that control the *degree* of such variability. For example, in [9, 22] the following Bayesian model is analyzed

$$Y_i|\theta_i \overset{ind}{\sim} \text{Poisson}(\theta_i E_i) \tag{1.11}$$

$$\theta_i|\eta \overset{ind}{\sim} \text{Gamma}(\alpha, v) \tag{1.12}$$

for $i = 1, 2, \ldots, N$, in which $\eta = (\alpha, v)'$ and a prior Gamma(α, v) for relative risks is introduced, whose marginal mean and variance are $E(\theta_i|\eta) = (\alpha/v)$ and $Var(\theta_i|\eta) = (\alpha/v^2)$. By integrating out random effects θ_i, simple calculations show that the marginal likelihood of the data is the product of N Negative Binomial densities each one having unconditional mean and variance given by

$$E(Y_i|\eta) = E_i \frac{\alpha}{v} \tag{1.13}$$

$$Var(Y_i|\eta) = E_i \frac{\alpha}{v} + E_i \frac{\alpha}{v^2} \tag{1.14}$$

from which it follows that the marginal areal variance is substantially larger than the mean. The posterior distribution of relative risks, assuming α and v to be known, is Gamma$(Y_i + \alpha, E_i + v)$. In other words, the posterior estimates of relative risks under a quadratic loss are given by

$$E(\theta_i|Y_i, \eta) = \frac{Y_i + \alpha}{E_i + v}. \tag{1.15}$$

A parametric empirical Bayesian (PEB) approach to relative risk estimation is proposed in [9], according to which unknown hyperparameters are estimated by numerically maximizing the joint marginal likelihood

$$h(\alpha, v) = \prod_{i=1}^{N} \int f(y_i|\theta_i)\pi(\theta_i|\eta) \, d\theta_i \tag{1.16}$$

and estimates $\hat{\alpha}$ and $\hat{v}$ are plugged into the marginal posterior expectations (1.15) to obtain the PEB estimate of θ_i as

$$\hat{\theta}_i^{(EB)} = \frac{Y_i + \hat{\alpha}}{E_i + \hat{v}} = \frac{E_i}{E_i + v}\hat{\theta}_i + \left(1 - \frac{E_i}{E_i + v}\right)\frac{\hat{\alpha}}{\hat{v}}. \tag{1.17}$$

Such an estimate has the form of a "shrinkage" estimator, being a weighted average of the area-specific SMR and the common prior mean. Each standardized mortality ratio $\hat{\theta}_i$ is pushed toward the global prior mean, and the size of this effect is proportionally larger in those areas where the number of expected events E_i is smaller. PEB techniques have a deep connection with James–Stein estimation, which has its roots in the Stein's proof that maximum likelihood estimation methods are inadmissible under the summed squared error loss beyond simple one- or two-dimensional situations [8]. Of course, it makes perfect sense to produce p-values-based maps in which an unconditional Negative-Binomial product-likelihood is used for calculating probabilities defined by expression (1.10).

Fully Bayesian estimation of relative risks considers various prior models for area-specific relative risk parameters that account for various aspects of extra-Poisson dispersion. For example, if no systematic spatial pattern to the variability of relative risks is present, a normal prior distribution on the logarithm of each relative risk is often used, as this leads to a generalized linear mixed models formulation that allows for the inclusion of area-specific covariate information. In the same way, we have often prior knowledge that geographically close areas tend to have similar relative risk; as in practice it is often unclear how to choose between an unstructured prior and a spatially structured prior, the following convolution model has been proposed by Besag et al. in a seminal paper [2]

$$\phi_i = \log(\theta_i) = \mu + \psi_i + \upsilon_i \tag{1.18}$$

where ψ_is allow for spatially structured risk patterns, being jointly defined by the following intrinsic GMRF [33]

$$\pi\left(\psi\,|\,\sigma_\psi^2\right) \propto \exp\left\{-\frac{1}{2\,\sigma_\psi^2}\sum_{i\sim j}\left(\psi_i - \psi_j\right)^2\right\} \tag{1.19}$$

which is the joint version of the conditional specification (1.5) and leads to an improper prior over the space of spatially structured effects. As we said before, a zero-mean multivariate Gaussian prior is commonly used for unstructured effects u_is

$$\pi\left(\upsilon\,|\,\sigma_\upsilon^2\right) \propto \exp\left\{-\frac{1}{2\,\sigma_\upsilon^2}\sum_{i=1}^{N} \upsilon_i^2\right\}. \tag{1.20}$$

The parameter μ is a baseline log-relative risk and the two prior components are assumed to be independent. Posterior log-relative risk estimates are smoothed in comparison with the SMRs, and probability maps may be drawn by estimating area-specific posterior probabilities $E[I(\theta_i>1)|Y] = \Pr\{\theta_i>1|Y\}$ (where $I(\cdot)$ denotes the event indicator function). Also for p-values-based maps, the resulting choroplets are likely to be insufficiently informative about the actual level of risk, but their aspect is often quite smoothed and they may be indeed useful to confirm the presence of "hotspot" of high-risk areas.

Of course, a Bayesian analysis may depend critically on the modeling assumptions because changes in the prior distributions may cause relevant changes in the posterior distributions. In fact, a crucial problem in the formulation of the BYM model is the specification of the prior distribution for the random effects variance parameters σ_ψ^2, σ_v^2 [6, 14]. As in the case of the less structured Poisson–Gamma model, these priors are parameterized by hyperparameters which control the variability of the relative risks across the map. One common choice is to specify independent inverse Gamma priors for σ_ψ^2, σ_v^2, but other elicitations are indeed possible: for example, [43] considers an inverse Gamma prior for the total variance $\sigma_T^2 = \sigma_\psi^2 + \sigma_v^2$ and a Beta prior for the spatial fraction

$$\text{SF} = p = \frac{\sigma_\psi^2}{\sigma_\psi^2 + \sigma_v^2} \tag{1.21}$$

which represents the proportion of the relative risk variation that is attributable to the spatial component: the closer the spatial fraction is to unity, the greater the relative risk posterior estimates are shrunken toward a local mean. By back-transforming from $\left(\sigma_T^2, p \right)$ to $\left(\sigma_\psi^2, \sigma_v^2 \right)$, it is possible to show that this specification induces positive dependence in the joint prior for $\left(\sigma_\psi^2, \sigma_v^2 \right)$. It is customary to evaluate the model sensitivity by choosing different prior distributions of the variance terms: some quite common alternatives will be compared in Sect. 1.2.4.

The computational machinery for calculating posterior parameter estimates has recently been enriched by a new arrival. Besides traditional (MCMC methods [10]), integrated nested Laplace approximation (INLA) provides a fast implementation of the Bayesian approach to generalized linear mixed models [34].

1.1.4 Identifiability Issues and Bayesian Model Choice

As we said before, the CAR prior (1.19) is not integrable, being equivalent to a joint multivariate distribution with singular covariance matrix, and hence corresponding to an improper prior. If an Uniform(R) improper prior is assumed for the overall log-relative risk μ [15], formally prove that the resulting posterior is not integrable, and Bayesian analysis will become impossible. Apart from a formal proof, a simple explanation may be provided for this: CAR prior only defines contrasts $\psi_i - \psi_j$ for $i \neq j$, but they do not identify an overall mean value for log-relative risks because they are translation invariant, and hence they confound the baseline effect. A common solution to generate a proper posterior is to identify the overall mean by adding the constraint

$$\sum_{i=1}^{N} \psi_i = 0 \tag{1.22}$$

which, in the terminology of [13], defines a *proper embedded posterior* (in the same paper [13], study in its full generality the conditions under which an embedded constrained lower-dimensional parameter has a unique proper posterior). It is interesting to note that the same constraint has to be added if we consider a BYM model with ecological covariates, for which the linear predictor has the form

$$\phi_i = \log(\theta_i) = \mu + x'_i\beta + \psi_i + \upsilon_i \qquad (1.23)$$

for a p-dimensional vector x_i of area-specific covariates and a vector of fixed coefficients β, for which an Uniform(R^p) improper prior is assumed.

If we have a set of q competing Bayesian specifications, a formal Bayesian procedure of model choice relies on *posterior model probabilities* [28]

$$P(M_j|y) \propto P(M_j)m_i(y), \quad j = 1, 2, \ldots, q \qquad (1.24)$$

where $P(M_j)$ is the prior model probability and $m_j(y)$ the *marginal likelihood* or prior predictive density

$$m_j(y) = \int f_i(y|\eta_j)\pi_i(\eta_j)\,\mathrm{d}\eta_j \qquad (1.25)$$

assuming that, under the model M_j, the data Y are assumed to have density $f_j\left(y|\eta_j\right)$ with prior distribution $\pi_j(\eta_j)$. With the Bayes maximum a posterior rule (BMAP), a sensible model choice procedure, on the basis of the only evidence provided by the data, would consider as the "best" model the specification M_s satisfying

$$P(M_s|y) = \max_j P\left(M_j|y\right), \quad j = 1, 2, \ldots, q. \qquad (1.26)$$

A particular common choice of prior model probabilities, often justified on the ground of an insufficient knowledge, is $P(M_j) = 1/q$ so that

$$P\left(M_j|y\right) \propto m_j(y), \quad j = 1, 2, \ldots, q. \qquad (1.27)$$

Apart from the fact that prior model probabilities seem often difficult to justify, harder difficulties arise when improper priors are set (as in the case of the BYM model). To understand why these difficulties are unavoidable, we consider a simpler notation for a problem involving the comparison of just two models. In this case the BMAP rule simply considers the ratio (we assume constant prior model probabilities)

$$B_{jk} = \frac{m_j(y)}{m_k(y)} = \frac{p_j/c_j}{p_k/c_k} = \frac{p_j}{p_k}\frac{c_k}{c_j} \qquad (1.28)$$

where B_{jk} is the Bayes factor in favor of model j against model k, and $\pi_j(\eta_j) \propto h_j(\eta_j)$ with normalizing constant being given by

$$c_j = \int h_j(\eta_j)\,\mathrm{d}\eta_j \qquad (1.29)$$

with identical meaning for c_k. In the same way, p_j is the unnormalized marginal likelihood

$$p_j = \int f_j\left(y|\eta_j\right) h_j(\eta_i)\, d\eta_j. \tag{1.30}$$

Of course, if a proper prior is used for each model such that $c_k < +\infty$ and $c_j < +\infty$ are well defined, the Bayes factor is well defined as the ratio c_k/c_j is also defined. When an improper prior is used such that $c_k = c_j = +\infty$, using a suitable limiting procedure [38] prove that the ratio c_k/c_j is either 0, 1, or $+\infty$ depending on the relative dimension of the two models. In particular, they show that if the parameter vector can be partitioned as $\eta_\ell = (\gamma_\ell, \gamma)'$, $\ell = j, k$, and improper priors of the same form are used only on γ, the Bayes factor is then well defined.

To what extent these results are useful to justify the comparison of several competing BYM models on the ground of the relative marginal likelihoods is still not fully understood. If we want rely on a sounder but more informal criterion [37], propose a generalization of the AIC based on the posterior distribution of the deviance statistics

$$D(\phi) = -2 \log f(y|\phi) + 2 \log h(y) \tag{1.31}$$

where the log-likelihood of the current model is compared to a baseline term $h(y)$ that is function of the data alone and hence does not affect posterior inference. For the Poisson likelihood (1.11), written in terms of area-specific log-relative risk exp $(\phi_i) = \theta_i$, the deviance statistics assume the following form

$$D(\phi) = 2\sum_{i=1}^{N} \left\{ Y_i \left[\log\left(\frac{Y_i}{\exp(\phi_i)E_i} \right) \right] - [Y_i - \exp(\phi_i)E_i] \right\} \tag{1.32}$$

provided that the standardized terms $h(y)$ is set equal to the saturated likelihood. The deviance information criterion (DIC) is defined as

$$\text{DIC} = \bar{D} + p_D \tag{1.33}$$

where $\bar{D}$ is the posterior expectation of the saturated deviance (1.31), and

$$p_D = \bar{D} - D[E(\phi|y)] \tag{1.34}$$

is the posterior expectation of deviance minus the deviance evaluated at the posterior expectations of log-relative risks. The proposed criterion is justified by Spiegelhalter et al. [37] by providing several arguments according to which $\bar{D}$ can be considered as a posterior summary of the goodness of fit of the actual model, whereas p_D is interpretable as a penalty term measuring the complexity of the model. This latter constant is commonly referred to as the *effective number of parameters*, in fact it is often less than the total number of parameters, due to the interdependencies across parameters in the likelihood introduced by random effects specified in higher levels (for example: the spatially structured random effect in the BYM model). It is clear that smaller values of the DIC will indicate a better-fitting model, after penalizing it for the complexity of the parameter space.

1.1.5 Local Cluster Detection: The Classical Spatial Scan Statistics

Beside the analysis of spatial variation in risk, spatial cluster detection is an important tool to identify areas of elevated risk and to generate hypotheses about disease etiology: as for disease mapping, there exists a considerable interplay between Bayesian and frequentist methods when local scanning of disease rates is the main interest. By definition, a *scanning window* Z_j is any collection of connected subareas in the study area (which will be denoted as G in this section) such that $z_j \cap z_k = \emptyset$ for $j \neq k$. Typically, a scanning window is a set of connected subareas whose centroids fall within a scanning circle. In any case, irregularly shaped zones following the area boundaries are possible. For notational simplicity, in this section we do not consider any subdivision into strata according to some confounding variable.

Let G denote the study area and $Z \subset G$ a generic scanning window such that X_Z and $X_{\bar{Z}}$ (with $\bar{Z} = G \backslash Z$) are independent nonhomogeneous Poisson point processes having intensities, respectively, given by

$$\lambda_Z(x) = p\mu(x)1_Z(x) \tag{1.35}$$

$$\lambda_{\bar{Z}}(x) = q\mu(x)1_{\bar{Z}}(x) \tag{1.36}$$

where p and q indicate the probability that one individual at risk, living, respectively, inside or outside the zone Z, has a given disease: the "background" spatial intensity $\mu(x)$ models the distribution of the population at risk over the area G. From these assumptions, it follows that

$$Y(Z) = \text{Poisson}(p\mu(Z)) \tag{1.37}$$

$$Y(\bar{Z}) = \text{Poisson}(q\mu(\bar{Z})) \tag{1.38}$$

where $Y(Z)$ represents the random number of cases falling within Z, with $Y(Z) \overset{\text{ind}}{\sim} Y(\bar{Z})$ (consequently, $Y(G) - Y(Z)$ represents the random number of events within $\bar{Z}$): to test the alternative hypothesis of raised incidence $H_1: p > q$ against the null $H_0: p = q$ [17, 18], consider a generalized likelihood-ratio test statistic which, under the Poisson likelihood, assumes the following form for a given subset Z

$$
\begin{aligned}
A_Z &= \frac{\max_{p>q} L(Z,p,q)}{\max_{p=q} L(Z,p,q)} \\
&\propto \left(\frac{Y(Z)}{E(Z)}\right)^{Y(Z)} \left(\frac{Y(G)-Y(Z)}{E(G)-E(Z)}\right)^{Y(G)-Y(Z)} I\left(\frac{Y(Z)}{E(Z)} > \frac{Y(G)-Y(Z)}{E(G)-E(Z)}\right)
\end{aligned}
\tag{1.39}
$$

where $E(Z) = p_R \mu(Z)$ is the expected number of cases within Z under a reference rate p_R, and $I(\cdot)$ is the indicator function.

Cluster detection is based on the test statistics

$$
\Lambda = \max_{Z \in \Delta} \Lambda_Z
\tag{1.40}
$$

for a suitable collection of scanning windows Δ. Assessing the statistical significance of (1.40) is a difficult problem, as the null sampling distribution is hard to derive. For these reasons, cluster inference commonly relies on Monte Carlo computation of approximated p-values, an approach which has been implemented in the SaTScan software [16]. In Monte Carlo hypothesis testing

$$
\text{simulated } p \text{ - value} = \frac{1 + \sum_{s=1}^{R} I\left(\Lambda^{(s)} \geq \Lambda_{obs}\right)}{1+R},
\tag{1.41}
$$

given that R datasets have been simulated independently under the null hypothesis and values of the test statistics $\left\{\Lambda^{(1)}, \Lambda^{(2)}, \ldots, \Lambda^{(R)}\right\}$ have been calculated accordingly. Secondary clusters with high likelihood value containing about the same areas are usually of little interest: more interestingly secondary clusters are those located in another part of the map and that do not overlap with the more likely one. Adaptation of the spatial scan statistics in order to detect disease clusters that occur in non-compact and non-circular shapes is another very deep problem that has received a lot of attention in the most recent literature [46].

1.1.6 A Fully Bayesian Approach to Scanning Spatial Disease Rates

When overdispersion is present, in most cases this is due to the fact that risk levels of areas that are close to each other will tend to be positively correlated, as they share a number of spatially varying characteristics. As a consequence, not only the Poisson assumption is violated but also excessive false alarms or type I errors will occur [21, 48]. For this reason [5] proposes a cluster detection algorithm based on the following variation of the standard BYM model ($i = 1, 2, \ldots, N$)

$$
Y_i | \theta_i, E_i \overset{ind}{\sim} \text{Poisson}(\theta_i E_i)
\tag{1.42}
$$

$$
\log(\theta_i) = \mu + \alpha_{Z_1} I[A_i \in Z_1] + \cdots + \alpha_{Z_s} I[A_i \in Z_s] + \psi_i + \upsilon_i
\tag{1.43}
$$

where, as in the previous section, Z_j is any collection of connected subareas (i.e., cluster) in the study area such that $Z_j \cap Z_k = \emptyset$ for $j \neq k$, and the cluster-specific fixed effect α_{Z_j} enters the model as the coefficient of indicator variable I $(A_i \in Z_j)$, which assumes value 1 if subarea A_i is in Z_j and 0 otherwise. The number s of non-overlapping clusters is assumed to be finite but unknown.

Prior specification in model space should keep into account, in some way, the non-overlapping constraint $Z_j \cap Z_k = \emptyset$. In other words, if we collect an initial set of candidate clusters/dummy variables as the columns of a matrix Δ whose rows represent sub-areas A_i, the model selection algorithm will have to select a subset of those columns in a suitable way, but not every possible subset corresponds to an admissible linear predictor in (1.43). Keeping this objective [5], formulate a sequential cluster detection algorithm based on the DIC criterion, which evolves according to the following steps:

Step 1. Suppose that the initial matrix $\Delta(1)$ has dimension $N \times \ell$, and let $\delta_j(1) = I$ $\left[A_i \in Z_j(1)\right]$, with $1 \leq i \leq N$, denote a single column of $\Delta(1)$. Conditionally on a fixed $\delta_j(1)$, we fit the single-cluster model given by

$$\log(\theta_i) = \mu + \alpha_{Z_j(1)} I\left[A_i \in Z_j(1)\right] + \psi_i + v_i \qquad (1.44)$$

for every $\delta_j(1), j = 1, 2, ..., \ell$. The whole collection of fitted models is put in order according to the DIC criterion (lower-value DICs indicate better-fitted models), and the first optimal cluster $\delta_{opt}(1)$ corresponds to the area-indicator variable entering that model having the lowest DIC value.

Step 2. Let $\Delta(2)$ be the current candidate cluster matrix. The columns of $\Delta(2)$ are a subset of those of $\Delta(1)$: we delete all those columns of $\Delta(1)$ corresponding to $\delta_{opt}(1)$ itself and to each scanning zone overlapping with the first optimal cluster. The current model is

$$\log(\theta_i) = \mu + \alpha_{Z_{opt}(1)} I\left[A_i \in Z_{opt}(1)\right] + \alpha_{Z_j(2)} I\left[A_i \in Z_j(2)\right] + \psi_i + v_i \qquad (1.45)$$

and the current set of scanning zones is assessed according to the DIC criterion by repeatedly fitting model (1.45) conditionally on the chosen $\delta_j(2)$. Whenever necessary, a second optimal cluster $\delta_{opt}(2)$ is identified (see Step k).

$$\vdots$$

Step k. The procedure stops at step k provided that it is not possible to improve data explanation by letting further cluster-specific terms enter the model: this eventuality can be easily assessed by means of the sequence of lowest DIC values of each previous step, in the sense that the algorithm stops when such sequence reaches its minimum and becomes increasing.

The final model contains $s = k$ fixed-effect terms

$$
\begin{aligned}
\log(\theta_i) \;=\; & \mu + \alpha_{Z_{\text{opt}}(1)} I\big[A_i \in Z_{\text{opt}}(1)\big] + \alpha_{Z_{\text{opt}}(2)} I\big[A_i \in Z_{\text{opt}}(2)\big] + \cdots + \\
& + \alpha_{Z_{\text{opt}}(k)} I\big[A_i \in Z_{\text{opt}}(k)\big] + \psi_i + \upsilon_i
\end{aligned}
\tag{1.46}
$$

corresponding to clusters satisfying the required constraint $Z_{\text{opt}}(g') \cap Z_{\text{opt}}(g'') = \emptyset$ for steps $g' \neq g''$, with $1 \leq g', g'' \leq k$.

The initial model space dimension of the fixed effect part, i.e. the $\Delta(1)$ matrix column space dimension, decays quite fast as new clusters are identified. However, [5] points out that despite the important theoretical developments that have been recorded in the last few years, the use of MCMC methods to estimate each model would be painfully slow from the end user's point of view, and suggest to use the newly developed INLA posterior integration numerical scheme [34], which requires far shorter computational times and makes the implementation of the proposed methodology computationally feasible. We defer to Sect. 1.2.5 for further elucidations concerning some practical aspects.

1.2 Case Study: Analysis of the Spatial Distribution of Liver Cancer in Apulia, Italy

1.2.1 Introduction

Liver cancer, also known as primary liver cancer or hepatoma, is a cancer arising from the liver. However, the term liver cancer can also refer to cancer that has spread to the liver from other organs. In this case, the disease is called metastatic or secondary. Liver cancer is one of the most common cancer in the world with a very low survival rate usually smaller than one year. The diseases strongly associated with liver cancer are chronic viral hepatitis, alcoholism, and cirrhosis. The role of each risk factor, as well as their interrelationship, is well established [19, 39, 47]. This cancer is highly frequent in Southeast Asia (China, Hong Kong, Taiwan, Korea, and Japan) and accounts for up to half of all cancers in some underdeveloped countries [27]. This is due to the prevalence of hepatitis B infection, which is usually a childhood disease and can be easily caught from contaminated blood or sexual contact. In these areas the cancer usually affects people 30–40 years old. In contrast, the cases of liver cancer in North America and Western Europe are much lower even though the rate of diagnosis is rising [20, 23]. This increase is due primarily to rising obesity and diabetes rates and to chronic hepatitis C. The alcohol abuse which causes cirrhosis is another very common cause of liver cancer in the developed countries. Here the people affected by this cancer are in their 60s and 70s and are men much more than women.

In Italy liver cancer is the seventh leading cause of cancer death even if the situation at the regional level is very heterogeneous. The spread of hepatocarcinoma in the Apulia region is substantial and it is related to hepatitis viruses that endemically affect the population. Currently the situation is stationary, but it is expected a reduction in the cancer incidence with the reduced incidence of hepatitis. In fact an improved prevention and in particular vaccination are favoring a constant descent of hepatitis B. On the other hand, a better public hygiene and many activities aimed at preventing the disease condition is driving a decrease in the number of hepatitis C cases. Particular attention is paid to liver cancer mortality in some areas of the Apulia showing a significant excess compared to the regional situation. The mortality rates for liver cancer in the period 2000–2005 are showed in the Causes of Death Atlas of the Apulia region [26]: in males, it detects the presence of a cluster of municipalities with high mortality, which belong to the northern area of the province of Bari and the newly constituted Barletta-Andria-Trani (BAT) province; in females, the geographical distribution of mortality is comparable to that described for males. In this study ample space will be given to further illustration of this aspect.

1.2.2 Data Source

The mortality data analyzed in this work are drawn from the Cause of Death Nominative Registry (RENCAM) of the Apulia region. Data are highly reliable as they are first collected by referents of smaller registries uniformly distributed over the whole region, then they are controlled, encoded, and compared with those of ISTAT (National Institute for Statistics). Data are publicly accessible and attached to the Causes of Death Atlas of the same region. Here, we consider the mortality cases, in both sexes, for liver and intrahepatic bile ducts malignant tumor (ICD-IX:155.0–155.1) occurring in the 258 municipalities of the Apulia during 2000–2005. The ICD-IX codes selected refer only to primary cancer. Age-standardized expected cases were obtained under indirect internal standardization: the number of person-years at risk was estimated by the 2000–2005 regional population (ISTAT source) divided into quinquennial age classes.

1.2.3 Some Homogeneity Tests

Before estimating parameters we test relative risks for the presence of heterogeneity, as the use of the BYM model needs to be motivated if the evidence for extra-Poisson variation is not strong. The heterogeneity may be of course related to spatially varying risk factors and may lead to an increased risk in the areas more exposed to these factors.

Testing heterogeneity may be based on traditional goodness-of-fit statistics, in which the null hypothesis to be tested is

$$H_0: Y_i|\theta \overset{ind}{\sim} \text{Poisson}(\theta E_i) \tag{1.47}$$

which is in turn equivalent to $H_0: \theta_1 = \theta_2 = \cdots = \theta_N = \theta$. Whenever θ is set a priori, the traditional chi-squared statistic

$$\chi^2 = \sum_{i=1}^{N} \frac{(Y_i - \theta E_i)^2}{\theta E_i} \tag{1.48}$$

follows, asymptotically and under the null hypothesis, a chi-squared distribution with N degrees of freedom. When θ is estimated from the data, given that the maximum likelihood estimator of θ in the restricted model (1.47) is $\hat{\theta} = \Sigma_i Y_i / \Sigma_i E_i$, it follows that (1.48) must be replaced by the test statistic

$$\chi^2 = \frac{\sum_{i=1}^{N} \left(Y_i - \frac{\Sigma_i Y_i}{\Sigma_i E_i} E_i\right)^2}{\frac{\Sigma_i Y_i}{\Sigma_i E_i} E_i}. \tag{1.49}$$

It is worth noting that under internal standardization $\hat{\theta} = 1$, and hence the null hypothesis corresponding to this special case is given by $H_0: \theta_1 = \theta_2 = \cdots = \theta_N = 1$. With the estimation of θ, one degree of freedom is lost, and the test statistic follows asymptotically a chi-squared distribution with $N - 1$ degrees of freedom [32]. Of course, it is straightforward and preferable to carry out a Monte Carlo test by randomly simulating counts data Y_i under the null hypothesis, and calculating the test statistic under each simulation. Comparison with the observed statistic leads easily to a Monte Carlo p-value computation: greater departures of counts Y_i from their null expected values θE_i produce greater χ^2 values.

The chi-squared test is expected to have reasonable power under many alternatives [31]: consider the uniformly most powerful test for the null (1.47) versus the alternative hypothesis that the relative risks θ_i are random effects drawn from the conjugate Gamma distribution. Specifically

$$H_1 : \theta_i | \lambda, \sigma^2 \overset{ind}{\sim} \text{Gamma}\left(\lambda^2/\sigma^2, \lambda/\sigma^2\right) \tag{1.50}$$

for $i = 1, 2, \ldots, N$, so that $E(\theta_i) = \lambda, Var(\theta_i) = \sigma^2$ and the counts follow an unconditional Negative-Binomial product likelihood. Unfortunately, [31] are able to derive a valid procedure only in the case in which θ is known. For the case in which θ is unknown, they exploit the results given in [30] for the Binomial–Multinomial case, using the fact that conditionally on $\Sigma_i Y_i$ the null model (1.47) becomes free of θ and is equivalent to

$$Y_1, Y_2, \ldots, Y_N \Big| \sum_i Y_i \sim \text{Multinomial}\left(\sum_i Y_i \{p_i; i = 1, 2, \ldots, N\}\right) \tag{1.51}$$

with area-specific probabilities $p_i = E_i / \Sigma_i E_i$, and test the null (1.51) against the alternative that counts are drawn from a conjugate Multinomial-Dirichlet random-effect model. The test statistic is

Table 1.1 Heterogeneity tests of relative risks: for males (M) and females (F), p-values are calculated using $R = 9999$ Monte Carlo replicated samples under the null sampling distribution indicated in the second column

Test	Sampling	p (M)	p (F)
Chi-squared	Ind. Poisson	0.0001	0.0001
PW	Multinomial	0.0001	0.0001
Tango ($\lambda = 30$)	Multinomial	0.0001	0.0001
Tango ($\lambda = 50$)	Multinomial	0.0001	0.0001
Tango ($\lambda = 100$)	Multinomial	0.0001	0.0001
Tango ($\lambda = 200$)	Multinomial	0.0001	0.0001

$$PW = \sum_{i=1}^{N} E_i \left[\sum_{i=1}^{N} \frac{Y_i(Y_i - 1)}{E_i} \right]. \tag{1.52}$$

Under H_0, (1.52) follows an asymptotic Normal distribution with mean $\Sigma_i Y_i$ $(\Sigma_i Y_i - 1)$ and variance $2(N - 1)(\Sigma_i Y_i - 1)$. Large values of PW indicate heterogeneity and, again, it is preferable to carry out a Monte Carlo test.

Another way to assess the conditional model (1.51) is once again via the Multinomial chi-squared goodness-of-fit statistic, to test whether area-specific counts Y_i deviate significantly from their expected values under the null area-specific disease occurrence probabilities $\{p_i; i = 1, 2, \ldots, N\}$. In this context [40] proposes a spatially modified chi-squared, in which the quadratic form measuring discrepancies between observed and expected cases is defined as

$$T = (r - p)' A(r - p) \tag{1.53}$$

where $p = (p_1, p_2, \ldots, p_N)'$, $r = (r_1, r_2, \ldots, r_N)'$ with $r_i = Y_i/\Sigma_i Y_i$ for $i = 1,$ $2, \ldots, N$. The matrix A measures the degree to which the areas are connected, being defined as $a_{ij} = \exp(-d_{ij}/\lambda)$ for $i \neq j$ and $a_{ii} = 1$ (d_{ij} is the distance between the centroids of areas i and j). Under the null hypothesis and conditionally on λ, [40] proves that (1.53) has an asymptotic chi-squared distribution, even though a finite-sample Monte Carlo test will be used in our example. Since the interpretation of the parameter λ is not clear, [41] suggests to repeat the procedure using different λ's and provides a method for facing with multiple testing problems.

The results of Monte Carlo tests applied to our data are shown in Table 1.1: in particular, the chi-squared and Tango statistics strongly support the evidence for very significant discrepancies between observed and expected counts under the homogeneity hypothesis. In the same way, the Potthoff–Whittinghill's test indicates that relative risk is widely varying, and their variation is not accounted for under a constant risk Poisson model. Non-spatial data analysis cannot therefore explain extra-Poisson variation in risk and must be supplemented by suitable spatial statistical analyses.

1.2.4 Spatial Data Analysis

If the null hypothesis (1.47) is valid, the expected number of events $E(Y_i|\theta)$ in area i is proportional to underlying population at risk. The weakness of chi-square goodness-of-fit and other global clustering statistics is that they do not take into account where the greatest deviations between observed and expected cases occur. For example, [36] points out that deviations from the null hypothesis may be either due to a pronounced cluster of regions where the fit is poor, or to elevated peaks within a more random spatial distribution. Methods of spatial analysis discussed in the previous section are therefore aimed at depicting the geographic variation in risk, and answering to the question whether the pattern seen is due to random fluctuations rather than to spatially varying etiologic risk factors.

Figure 1.1 shows a map of SMRs for liver cancer mortality (separately for both sexes). Areas were classified into gray levels according to the quintiles of the empirical distribution. For males, SMRs are widely varying around their overall mean of 0.90, ranging from 0 to 4.24 with standard deviation of 0.63; for females, the variation is comparable, ranging from 0 to 3.35 with an overall mean of 0.83 and a standard deviation of 0.48. These results further support the rejection of the null hypothesis of homogeneity of relative risks. Indeed, in agreement with the Potthoff–Whittinghill test, they suggest the presence of spatially varying deviations from a random pattern. Some additional information comes from the probability maps represented in Fig. 1.1, obtained by subdividing estimates of $(1 - \tilde{\rho}_i)$ probabilities into ten equidistant gray levels (areas with higher risks are those shown in the darkest grays). Despite the disturbing level of sampling noise, it is apparent that a cluster of neighboring areas at high risk is found for both sexes (although to a lesser extent for women). This cluster involves a group of municipalities in the northern of the Bari area, and touches areas falling inside the BAT province.

MCMC methods were used to obtain fully Bayesian estimates of the relative risks under the BYM model. The following conjugate Gamma priors for inverses of variances (that is precisions $\tau_\psi^2 = \sigma_\psi^{-2}$ and $\tau_\upsilon^2 = \sigma_\upsilon^{-2}$) and Uniform priors for variances were compared:

1. $\tau_\psi^2 \sim \text{Gamma}(0.5, 0.0005)$, $\tau_\upsilon^2 \sim \text{Gamma}(0.5, 0.0005)$.
2. $\tau_\psi^2 \sim \text{Gamma}(1.58 \times 10^{-5}, 3.98 \times 10^{-5})$, $\tau_\upsilon^2 \sim \text{Gamma}(4.42 \times 10^{-4}, 2.10 \times 10^{-4})$
   for males;

   $\tau_\psi^2 \sim \text{Gamma}(3.73 \times 10^{-6}, 1.93 \times 10^{-5})$, $\tau_\upsilon^2 \sim \text{Gamma}(1.04 \times 10^{-4}, 1.02 \times 10^{-4})$

   for females.
3. $\tau_\psi^2 \sim \text{Gamma}(0.1, 0.1)$, $\tau_\upsilon^2 \sim \text{Gamma}(0.001, 0.001)$.
4. $\tau_\psi^2 \sim \text{Gamma}(0.1, 0.01)$, $\tau_\upsilon^2 \sim \text{Gamma}(0.1, 0.01)$.
5. $\sigma_\psi^2 \sim \text{Uniform}(0, 1)$, $\sigma_\upsilon^2 \sim \text{Uniform}(0, 1)$.
6. $\sigma_\psi^2 \sim \text{Uniform}(0, 100)$, $\sigma_\upsilon^2 \sim \text{Uniform}(0, 100)$.

Given that no source of information relevant to our study was available, a range of non-informative priors were used except in the case of model 2, for which a weakly informative data-based prior was set, separately for both sexes, along the lines of [24]. In synthesis, location and scale parameters of inverse variances

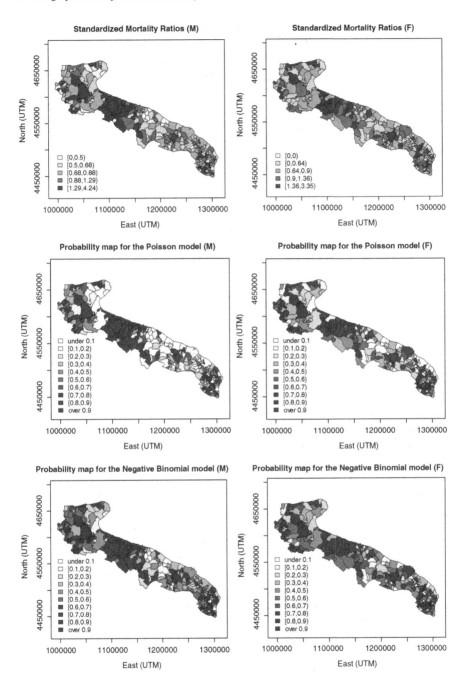

Fig. 1.1 From *top* to *bottom*: maximum likelihood estimates (SMRs) of relative risk of the mortality from liver cancer in Apulia, Italy, 2000–2005, and the corresponding probability maps under the product-Poisson and the product Negative-Binomial likelihood

Table 1.2 Sensitivity analyses for spatial Bayesian analysis of liver cancer mortality among males

Smoothed SMRs	Prior 1	Prior 2	Prior 3	Prior 4	Prior 5	Prior 6
Mean	93.00	92.90	92.90	92.70	92.80	92.70
Standard deviation	23.40	23.80	23.90	24.00	24.40	24.40
Maximum	262.80	263.30	263.40	263.40	264.70	263.90
75 % quartile	104.40	103.90	103.60	103.00	103.50	102.50
Median	83.00	83.60	83.50	83.50	83.40	83.30
25 % quartile	72.60	72.20	72.60	72.20	71.80	71.80
Minimum	44.90	44.60	44.60	44.80	44.10	44.10
90 % ratio	2.65	2.70	2.71	2.69	2.73	2.74
p_D	88.70	91.70	91.90	93.30	96.00	95.90
DIC	1193.20	1193.90	1193.10	1193.80	1193.60	1193.60
Spatial fraction	0.98	0.94	0.94	0.89	0.89	0.89

Gamma hyperpriors were chosen in such a way that the prior means were, respectively, set equal to

$$2 \times \left(\frac{1.65}{a}\right)^2, \quad \text{for } \tau_v^2 \tag{1.54}$$

$$\frac{2}{\bar{m}} \times \left(\frac{1.15}{a}\right)^2, \quad \text{for } \tau_\psi^2 \tag{1.55}$$

where $\bar{m}$ is the average number of neighbors across the study area, and the symmetric interval (e^{-a}, e^a) contains about of 90 % of SMRs for males and 75 % for females. Prior variances were set equal to 10^4 to reflect large uncertainty about the values set for prior means.

To formally compare the six models and to identify the one that fits better our data, we computed the (DIC, see Sect. 1.1.4), according to which a model with a smaller values of DIC is to be preferred. Several posterior summaries are shown in Tables 1.2 and 1.3 in correspondence of each proposed prior, including the spatial fraction (1.21) as well as the 90 % relative risk ratio, defined as the mean of the posterior distribution of the ratio of the 95th to 5th percentile of relative risks. The posterior means of variances of both the structured and the unstructured component are very similar in each analysis, but the variance of the spatial effect dominates: the spatial fraction is always greater than 60 % for females, a value that is as high as 89 % for men. These values are strongly compatible with the presence of a spatially structured heterogeneity of the relative risks.

Fully Bayesian estimates of posterior probabilities $\Pr\{\theta_i > 1 | Y\}$ for the "best" model (respectively, prior three for males and prior five for females) are shown in Fig. 1.2. The variation is much less pronounced than that seen with the probability maps based on the product-Poisson or product Negative-Binomial likelihood. The spatial structure is always dominated by the large cluster present in the north of

Table 1.3 Sensitivity analyses for spatial Bayesian analysis of liver cancer mortality among females

Smoothed SMRs	Prior 1	Prior 2	Prior 3	Prior 4	Prior 5	Prior 6
Mean	92.80	92.60	92.40	92.50	92.10	92.20
Standard deviation	24.80	25.30	26.10	25.80	26.80	26.70
Maximum	196.60	197.50	197.80	196.40	199.70	198.80
75 % quartile	101.00	101.60	100.80	100.60	100.90	100.40
Median	86.90	85.80	86.10	86.30	85.40	85.60
25 % quartile	79.90	78.90	78.50	78.90	77.30	77.80
Minimum	58.40	57.60	57.80	58.70	57.10	57.70
90 % ratio	1.90	1.94	1.93	1.87	1.97	1.93
p_D	61.40	63.90	67.30	66.40	71.00	70.70
DIC	1002.00	1002.30	1001.30	1001.20	1000.10	1001.30
Spatial fraction	0.87	0.88	0.82	0.77	0.79	0.75

Bari, although areas near the northwestern border plays now a significant role for men. In this latter case we cannot assess the size of edge effects, as it is well known that many analyses can be greatly altered by the inclusion of edge information [42].

As we said before, probability maps are not informative about the true level of risk: there will always be some geographic patterns apparent to the naked eye, and even in a fully Bayesian context it is important to assess whether these results are due to chance or not. This explains why it is useful to supplement the information provided in the maps by a graph showing the ranked fully Bayesian relative risk estimates for each area with the associated 95 % posterior credible intervals (Fig. 1.2). Higher relative risks are associated with credible intervals not including unity, strongly confirming the evidence of geographic variation due to a cluster of areas with significantly more cases than expected (an effect that is unlikely to be due to sampling noise under the BYM Bayesian framework). Of course, this conclusion follows from the *combined* information provided by Fig. 1.2.

1.2.5 Local Cluster Detection

As we said before in Sect. 1.1.5, cluster detection tests are concerned with local cluster detection: they are used when there is a simultaneous interest in testing the statistical significance of high-risk areas as well as detecting their location. The Kulldorff–Nagarwalla (KN) scan statistic is usually the first choice: for males and females, the most likely statistically significant clusters are reported in Fig. 1.3. Circles are constructed so that only those that contain up to 10 % of the total expected cases are considered. All calculations involving maximum-likelihood ratio scan statistic were performed using the SaTScan™ software,[1] version 9.1.1 Under the product-Poisson likelihood, a significant primary cluster is present for both males and females ($p = 0.001$), so that the presence of a large cluster located in the north of Bari is definitely confirmed.

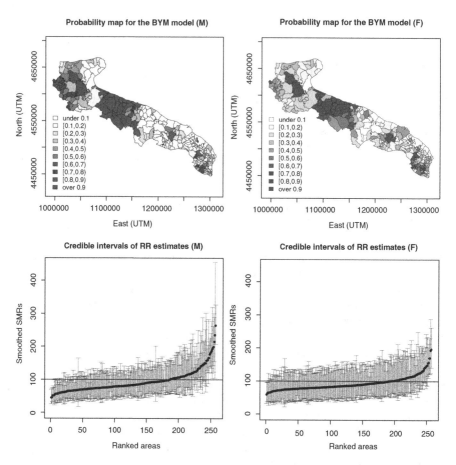

Fig. 1.2 *Top*: probability maps under the BYM model of the risk of liver cancer mortality in Apulia, Italy, 2000–2005. *Bottom*: ranked smoothed relative risks for each area, with the associated 95 % posterior credible intervals

The results of the Bayesian model-based cluster detection algorithm described in Sect. 1.1.6 was obtained using version 1.2b of the software written by the developers of the method [5]. In synthesis, the algorithm implementation focuses on creating the matrix whose column encodes all candidate clusters and on posterior parameter estimation and optimal cluster searches using INLA integration facilities to make the algorithm of Sect. 1.1.6 computationally feasible. The final optimal model is elaborated iteratively by removing all those clusters whose associated posterior credibility intervals include the zero log-relative risks. There are several available strategies for defining the initial collection of clusters, even though in this example we construct a sequence of circles of increasing radii from zero to the maximum distance between two areas, the total number of circles being set via the n.circles parameter. A scanning window includes all those areas whose centroid falls within a given circle, provided that the associated number of expected

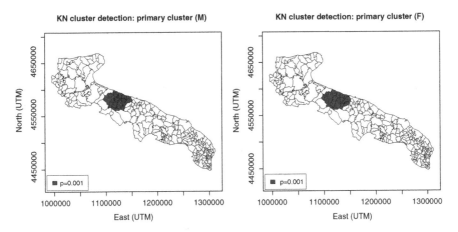

Fig. 1.3 Primary (most likely) clusters found by the Kulldorff–Nagarwall's spatial scan statistics, for the risk of liver cancer mortality in Apulia, Italy, 2000–2005

cases must not exceed 10 % of the total number of expected cases (of course, the value of this fraction is a fully tunable parameter).

After removing those clusters whose 95 % posterior credible intervals include zero, several decision rules are available to decide which of them should be considered as the primary and secondary zone where raised incidence/mortality occurs. A simulation study is provided in [5], in which the relative merits and demerits of some possible choices are examined. Here, we consider the uncertainty associated with posterior estimates, in the sense that cluster terms are ranked according to the length, from shortest to longest, of 95 % posterior credible interval of the associated cluster-specific effect α_{Z_j}. Simulation results shown in [5] indicate that this Bayesian cluster detection procedure outperforms traditional KN statistic in terms of correctly classified areas, even when a product Negative-Binomial likelihood is considered. The results of primary cluster detection are shown in Fig. 1.4, and no remarkable differences with the KN statistic output are found. These indications may be considered worthy of further investigation and may lead to the generation of new hypotheses concerning the origin of the disease.

1.2.6 Linking Mortality, Exposure and Hazards

The knowledge of socioeconomic factors is sometimes crucial to describe social inequalities in health. In a disease mapping framework and in absence of individual data, ecological or contextual measures of socioeconomic level are frequently used, so we give some illustration of the association between liver cancer mortality and material deprivation indicators. The hypothesis is strengthened by the existence, well documented in literature, of inverse gradients between the socioeconomic status (SES) and the mortality rates of numerous health outcomes. For our purposes

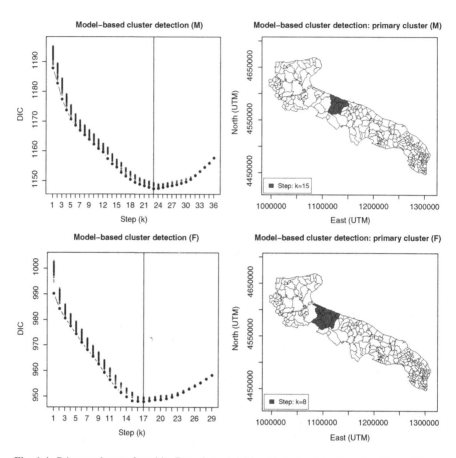

Fig. 1.4 Primary clusters found by Bayesian model-based cluster detection algorithm, with $n.$ $\texttt{circles} = 100$, for the risk of liver cancer mortality in Apulia, Italy, 2000–2005

a deprivation index at the municipal level will be compared, separately for both sexes, with the smoothed SMR estimates. The deprivation index is developed by Caranci et al. [7] for the whole national area and also separately for each Italian region. The demographic and socioeconomic data come from the 2001 National Population Census, conducted by the National Institute for Statistics (ISTAT). To characterize accurately the deprivation level, the authors select five socioeconomic and demographic variables reflecting the multiple dimensions of deprivation: education attainment, job, housing characteristics, and family structure. More specifically, the five selected indicators consist of 4 % and a ratio:

1. X_1: % of less educated people.
2. X_2: % of unemployed people.
3. X_3: % of homes for rent.
4. X_4: % of single-parent families.
5. X_5: household crowding index (number of co-residents per 100 m^2).

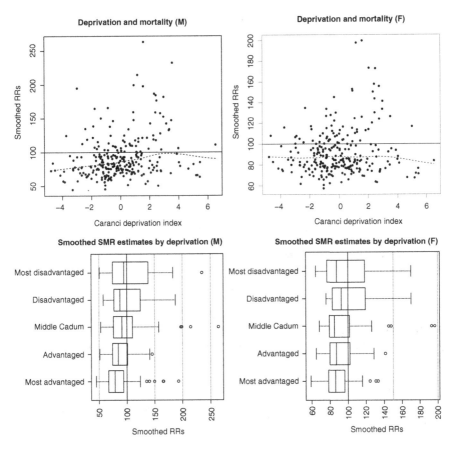

Fig. 1.5 Association between an area-specific deprivation measure and the smoothed risk of liver cancer mortality in Apulia, Italy, 2000–2005

The variables are transformed into Z-scores to center the distribution around zero and exclude the effect of their different variability

$$Z_i = \frac{X_i - \mu_{X_i}}{\sigma_{X_i}} \tag{1.56}$$

where μ_{X_i} and σ_{X_i} are the regional (Apulia) mean and standard deviation, respectively. The final index is the sum of standardized scores

$$index = \sum_{i=1}^{5} Z_i \tag{1.57}$$

and is classified according to the quintiles of the population. The upper part of Fig. 1.5 shows the relationship between the deprivation index and the smoothed

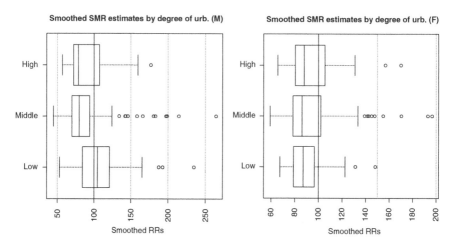

Fig. 1.6 Association between an area-specific degree-of-urbanization measure and the smoothed risk of liver cancer mortality in Apulia, Italy, 2000–2005

SMRs, but it is not possible to identify a strong correlation between variables both for men and for women. The pattern of dots is dispersed and without a clear slope. Some more information can be detected by observing the lower part of Fig. 1.5: especially for men, higher mortality risk for liver cancer is observed in the category "most disadvantaged" even if the high variability could affect the interpretation.

The fact that the distribution of relative risks of mortality for liver cancer exhibits a strong spatial structure might be explained in some other more suitable ways: the differentials on mortality could be associated with other factors besides the material deprivation. For example, urbanization may have an impact on population health because it brings social and economic changes that can raise risk factors associated with chronic disease. In other words, the possible relation between the degree of urbanization and the smoothed SMR estimates is another important aspect that must be kept into account. The urbanization index is extracted from the database of 2001 Istat National Population Census and has three levels: high (the most densely populated municipalities), medium, and low.

The municipalities less urbanized are those with a higher risk of mortality for males (Fig. 1.6), whereas no strong hypothesis of association between degree of urbanization and mortality for liver cancer could be done for females. Surprisingly, the areas at high risk in our analysis (the northern of the Bari area and the municipalities belonging to the BAT province) are urbanized and advantaged. This leads to new research directions to identify other risk factors. Linking mortality with socioeconomic variables does not help in this case also because there are different results between males and females. The attention could be focused on the occupational factors which expose the two sexes differently to the risk of liver cancer. For example, the exposure to vinyl chloride, a chemical used in making some kinds of plastics, raises the risk of hepatocellular cancer. This is only a suggestion as the scope of this work is to provide an overview of the methods used to analyze the spatial distribution of a disease and not to investigate its etiology.

1.3   Conclusions

This paper reviews some methods in the geographical analysis of diseases. Our focus was on Bayesian methods, which allow statistical inference in complex models for spatial dependence. We also discussed a number of issues relating to Bayesian spatial disease modeling, for example the sensitivity of posterior inference to prior setup, as well as mathematical difficulties inherent to formal Bayesian methods for model comparison when applied to spatial models for aggregated data. Of course, disease mapping studies are of an observational type, and we must be very careful to the danger of misinterpretation of the results, especially when the confounding effect of area-specific ecological variables such as deprivation and socioeconomic status are examined. Building an aggregate model from the individual model is an important step towards an understanding of the disease etiology, but potential sources of ecological bias should always be suspected. The long latency of some pathologies, and the migration problem is another important issue that could lead to misleading interpretations.

We also considered the problem of identifying disease clusters of neighboring areas sharing a raised disease risk. With the help of newly designed INLA numerical techniques for Bayesian posterior computations, we were able to provide a new method for scanning spatial disease which provide valid inferences when overdispersion is present, due to the fact that risk levels of neighboring areas will tend to be positively correlated as they share a number of spatially varying characteristics. In conclusion, we found a great appeal for the application of Bayesian approaches to the analysis of aggregated spatial data, and the development of those ideas will likely result in exciting advancements over the next years in the field of spatial epidemiology.

References

1. Banerjee, S., Carlin, B.P., Gelfand, A.E.: Hierarchical Modeling and Analysis for Spatial Data. Chapman & Hall/CRC, Boca Raton (2004). ISBN 978-0-203-48780-8
2. Besag, J.E., York, J.C., Mollié, A.: Bayesian image restoration with two applications in spatial statistics (with discussion). Ann. Inst. Stat. Math. **43**, 1–59 (1991). doi:10.1007/BF00116466
3. Bernardinelli, L., Pascutto, C., Best, N.G., Gilks, W.R.: Disease mapping with errors in covariates. Stat. Med. **16**, 741–752 (1997). doi:10.1002/(SICI)1097-0258(19970415) 16:7<741::AID-SIM501>3.0.CO;2-1
4. Besag, J., Kooperberg, C.: On conditional and intrinsic autoregression. Biometrika **82**, 733–746 (1995). http://www.jstor.org/stable/2337341
5. Bilancia, M., Demarinis, G.: Bayesian scanning of spatial disease rates with INLA (submitted)

6. Browne, W., Draper, D.: A comparison of Bayesian and likelihood-based methods for fitting multilevel models. Bayesian Anal. **1**(3), 473–514 (2006). doi:10.1214/06-BA117
7. Caranci, N., Costa, G.: Un indice di deprivazione a livello aggregato da utilizzare su scala nazionale: giustificazioni e composizione dell''indice. In: Costa, G., Cislaghi, C., Caranci, N. (eds) Disuguaglianze sociali di salute. Problemi di definizione e di misura. "Salute e Società" VIII, 1 (2009). doi:10.3280/SES2009-001006
8. Carlin, B.P., Louis, T.A.: Bayesian Methods for Data Analysis. Chapman & Hall/CRC, Boca Raton (2008). ISBN 978-1-58488-697-6
9. Clayton, D., Kaldor, J.: Empirical Bayes estimates of age-standardized relative risks for use in disease mapping. Biometrics **43**(3), 671–681 (1987). http://www.jstor.org/stable/253200
10. Cowles, M.K., Carlin, B.P.: Markov Chain Monte Carlo convergence diagnostics: a comparative review. J. Am. Stat. Assoc. **91**, 883–904 (1996). http://www.jstor.org/stable/2291683
11. Cramb, S., Mengersen, K.L., Baade, P.: Developing the atlas of cancer in Queensland: methodological issues. Int. J. Health Geogr. **10**, 9 (2011). doi:10.1186/1476-072X-10-9
12. Dabney, A.R., Wakefiled, J.C.: Issues in the mapping of two disease. Stat. Methods Med. Res. **14**, 1–30 (2005). doi:10.1191/0962280205sm340oa
13. Gelfand, A.E., Sahu, S.K.: Identifiability, improper priors and Gibbs sampling for generalized linear models. JASA **94**, 247–253 (1999). http://www.jstor.org/stable/2669699
14. Gelman, A.: Prior distributions for variance parameters in hierarchical models (comment on article by Browne and Draper). Bayesian Anal. **1**(3), 515–534 (2006). doi:10.1214/06-BA117A
15. Ghosh, M., Natarajan, K., Waller, L.A., Kin, D.: Hierarchical Bayes GLMs for the analysis of spatial data: an application to disease mapping. J. Stat. Plan. Inference **75**(2), 305–318 (1999). doi:10.1016/S0378-3758(98)00150-5
16. Gómez-Rubio, V., Ferrándiz-Ferragud, J., López-Quílez, A.: Detecting clusters of disease with R. J. Geogr. Syst. **7**, 189–206 (2005). doi:10.1007/s10109-005-0156-5
17. Kulldorff, M., Nagarwalla, N.: Spatial disease clusters: detection and inference. Stat. Med. **14**, 799–810 (1995). doi:10.1002/sim.4780140809
18. Kulldorff, M.: A spatial scan statistic. Commun. Stat. Theory Methods **26**(6), 1481–1496 (1997). doi:10.1080/03610929708831995
19. La Vecchia, C., Negri, E., Decarli, A., et al.: Risk factors for hepatocellular carcinoma in Northern Italy. Int. J. Cancer **42**, 872–876 (1988)
20. Levi, F., Lucchini, F., Negri, E., Boyle, P., La Vecchia, C.: Cancer mortality in Europe, 1995–1999, and an overview of trends since 1960. Int. J. Cancer **110**, 155–169 (2004). doi:10.1002/ijc.20097
21. Loh, J.M., Zhu, Z.: Accounting for spatial correlation in the scan statistics. Ann. Appl. Stat. **1**(2), 560–584 (2007). doi:10.1214/07-AOAS129
22. Marshall, R.J.: Mapping disease and mortality rates using empirical Bayes estimators. J. R. Stat. Soc. C Appl. Stat. **40**(2), 283–294 (1991). http://www.jstor.org/stable/2347593
23. McGlynn, K.A., Tsao, L., Hsing, A.W., Devesa, S.S., Fraumeni Jr., J.F.: International trends and patterns of primary liver cancer. Int. J. Cancer **94**, 290–296 (2001). doi:10.1002/ijc.1456
24. Mollié, A.: Bayesian mapping of Hodgkin's disease in France. In: Elliot, P., Wakefield, J., Best, N., Briggs, D. (eds.) Spatial Epidemiology Methods and Applications, pp. 267–285. Oxford University Press, Oxford (2001). ISBN 978-0198515326
25. Ocaña-Riola, R.: Common errors in disease mapping. Geospat. Health **4**(2), 139–154 (2010). http://www.geospatialhealth.unina.it/articles/v4i2/gh-v4i2-02-ocana-riola.pdf
26. Osservatorio Epidemiologico Regionale Puglia: Atlante delle cause di Morte della Regione Puglia. Anni 2000–2005 (2008). http://www.oerpuglia.org/Atlante.asp
27. Parkin, D.M., Bray, F., Ferlay, J., Pisani, P.: Estimating the world cancer burden: Globocan 2000. Int. J. Cancer **94**, 153–156 (2001). doi:10.1002/ijc.1440
28. Pericchi, L.R.: Model selection and hypothesis testing based on objective probabilities and Bayes factors. In: Dey, D.K., Rao, C.R. (eds.) Handbook of Statistics 25. Bayesian Thinking: Modeling and Computation, pp. 115–149. Amsterdam, Elsevier (2005). ISBN 9780444515391

29. Pfeiffer, D.U., Robinson, T.U., Stevenson, M., Stevens, K.B., Rogers, D.J., Clements, A.C.A.: Spatial Analysis in Epidemiology. Oxford University Press, Oxford (2008). ISBN 978-0-19-85098-82

30. Potthoff, R.F., Whittinghill, M.: Testing for homogeneity: I. The binomial and multinomial distribution. Biometrika **53**, 167–182 (1966). http://www.jstor.org/stable/2334062

31. Potthoff, R.F., Whittinghill, M.: Testing for homogeneity: II. The Poisson distribution. Biometrika **53**, 183–190 (1966). http://www.jstor.org/stable/2334063

32. Rao, C.R.: Advanced Statistical Methods in Biometric Research. Wiley, New York (1952)

33. Rue, H., Held, L.: Gaussian Markov Random Fields. Theory and Applications. Chapman & Hall/CRC, Boca Raton (2005). ISBN 978-1-58488-432-3

34. Rue, H., Martino, S., Chopin, N.: Approximate Bayesian inference for latent Gaussian models using Integrated Nested Laplace approximations (with discussion). J. R. Stat. Soc. B **71**, 319–392 (2009). doi:10.1111/j.1467-9868.2008.00700.x

35. Richardson, S., Guihenneuc, C., Lasserre, V.: Spatial linear models with autocorrelated error structure. Statistician **41**, 539–557 (1992). http://www.jstor.org/stable/2348920

36. Rogerson, P.A.: The detection of clusters using a spatial version of the chi-square goodness-of-fit statistic. Geogr. Anal. **31**(1), 130–147 (1999). doi:10.1111/j.1538-4632.1999.tb00973.x

37. Spiegelhalter, D.J., Best, N.G., Carlin, B.P., Linde, A.: Bayesian measures of model complexity and fit (with discussion). J. R. Stat. Soc. B **64**, 583–639 (2002). doi:10.1111/1467-9868.00353

38. Strachan, R.W., van Dijk, H.K.: Divergent priors with well defined Bayes factors. Tinbergen Institute Discussion Paper, TI 2011-006/4 (2011). http://papers.ssrn.com/sol3/DisplayAbstractSearch.cfm

39. Tabor, E., Kobayashi, K.: Hepatitis C virus, a causative infectious agent of non-A, non-B hepatitis: prevalence and structure – summary of a Conference on Hepatitis C virus as a cause of hepatocellular carcinoma. J. Natl. Cancer Inst. **84**, 86–90 (1992). doi:10.1093/jnci/84.2.86

40. Tango, T.: A class of test for detecting 'general' and 'focused' clustering of rare diseases. Stat. Med. **14**, 2323–2334 (1995). doi:10.1002/sim.4780142105

41. Tango, T.: A test for spatial disease clustering adjusted for multiple testing. Stat. Med. **19**, 191–204 (1995). doi:10.1002/(SICI)1097-0258(20000130)19:2<191::AID-SIM281>3.0. CO;2-Q

42. Vidal Roidero, C.L., Lawson, A.B..: An evaluation of edge effects in disease map modeling. Comput. Stat. Data Anal. **49**, 45–62 (2005). doi:10.1016/j.csda.2004.05.012

43. Wakefield, J.: Disease mapping and spatial regression with count data. Biostatistics **8**(2), 158–183 (2007). doi:10.1093/biostatistics/kxl008

44. Waller, L.A., Carlin, B.P., Xia, H., Gelfand, A.E.: Hierarchical spatiotemporal mapping of disease rates. J. Am. Stat. Assoc. **92**, 607–617 (1997). http://www.jstor.org/stable/10.2307/2965708

45. Waller, L.A., Gotway, C.A.: Applied Spatial Statistics for Public Health Data. Wiley, New Jersey (2004). ISBN 0-471-38771-1

46. Yiannakoulias, N., Rosychuk, R.J., Hodgson, J.: Adaptations for finding irregularly shaped disease clusters. Int. J. Health Geogr. **6**, 28 (2007). doi:10.1186/1476-072X-6-28

47. Yu, M.C., Mack, T., Hanisch, R., et al.: Hepatitis, alcohol consumption, cigarette smoking, and hepatocellular carcinoma in Los Angeles. Cancer Res. **43**, 6077–6079 (1983). http://cancerres.aacrjournals.org/content/43/12_Part_1/6077.full.pdf

48. Zhang, T., Zhang, Z., Lin, G.: Spatial scan statistics with overdispersion. Stat. Med. 13;**31**(8): 762–774 (2012). doi:10.1002/sim.4404

Chapter 2
A Fuzzy Approach to Ward's Method of Classification: An Application Case to the Italian University System

Francesco Campobasso and Annarita Fanizzi

Abstract A great part of statistical techniques has been thought for exact numerical data, although available information is often imprecise, partial, or not expressed in truly numerical terms. In these cases the use of fuzzy numbers can be seen as an appropriate way for a more effective representation of observed data. Diamond introduced a metrics into the space of triangular fuzzy numbers in the context of a simple linear regression model; in this work we suggest a multivariate generalization of such a distance between trapezoidal fuzzy numbers to be used in clustering techniques. As an application case of the proposed measure of dissimilarity, we identify homogeneous groups of Italian universities according to graduates' opinion (itself fuzzy) on many aspects concerning internship activities, by disciplinary area of teaching. Since such an opinion depends not only on the quality of internships, but also on the local context within which the activity is carried out, the obtained clusters are analyzed paying attention particularly to the membership of each university to Northern, Central, or Southern Italy. [This work is the result of joint reflections by the authors, with the following contributions attributed to Campobasso (Sects. 2.2, 2.3.2 and 2.4), and to Fanizzi (Sects. 2.1, 2.3 and 2.3.1).]

Keywords Diamond's distance • Internship activities • Italian universities • Trapezoidal fuzzy numbers • Ward's method

F. Campobasso (✉) • A. Fanizzi
Department of Economics and Mathematics, University of Bari, Via C. Rosalba 53, 70100 Bari, Italy
e-mail: francesco.campobasso@uniba.it; a.fanizzi@dss.uniba.it

S. Montrone and P. Perchinunno (eds.), *Statistical Methods for Spatial Planning and Monitoring*, Contributions to Statistics, DOI 10.1007/978-88-470-2751-0_2, © Springer-Verlag Italia 2013

2.1   Introduction

In most of empirical contexts the data obtained for decision making are only approximately known [1].

Modalities of quantitative variables are usually given as single exact numbers, although there are several sources of imprecision and uncertainty in their measurement, which prevent a researcher from obtaining the corresponding effective values; categories of qualitative variables are verbal labels of sets with vague borders and are often expressed in quantitative terms, which represent only accumulation values on an ideal *continuum* along which such categories are distributed.

Moreover usual descriptive statistics provide important indications about the intensity of a certain phenomenon, but they generate an unavoidable loss of information on the collected data. For example the mean represents an ideal summary value even if it ignores the variability of the observed population.

In 1965 Zadeh [12] introduced the fuzzy set theory in order to manage more appropriately phenomenons with uncertain borders [2, 11]; in this work we use trapezoidal fuzzy numbers, which allow us to obtain a more informative synthesis of the collected data.

2.2   A Generalization of Diamond's Distance

Diamond [9] introduced a metric into the space of triangular fuzzy numbers. A triangular fuzzy number $\tilde{X} = (x, x_L, x_R)_T$ for the variable X is characterized by a function $\mu_{\tilde{X}} : X \to [0, 1]$, like the one represented in Fig. 2.1, that expresses the membership degree of any possible value of X to $\tilde{X}$.

The accumulation x value is considered the core of the fuzzy number, while $\underline{\xi} = x - x_L$ and $\bar{\xi} = x_R - x$ are considered the left spread and the right spread, respectively. Note that x belongs to $\tilde{X}$ with the highest degree (equal to 1), while the other values included between the left extreme x_L and the right extreme x_R belong to $\tilde{X}$ with a gradually lower degree.

According to Diamond's metric, the distance between $\tilde{X}$ and $\tilde{Y}$ is:

$$d(\tilde{X}, \tilde{Y})^2 = d\big((x, x_L, x_R)_T, (y, y_L, y_R)_T\big)^2 = (x - y)^2 + (x_L - y_L)^2 + (x_R - y_R)^2.$$

A generalization of such a distance to trapezoidal fuzzy numbers in a multidimensional context is now used as a dissimilarity measure for hierarchical cluster techniques.

A trapezoidal fuzzy number $\tilde{X} = (x_1, x_2, x_L, x_R)_T$ for the variable X is characterized by a function $\mu_{\tilde{X}} : X \to [0, 1]$, like the one represented in Fig. 2.2, that expresses the membership degree of any possible value of X to $\tilde{X}$ and fits to represent statements like "between x_1 and x_2" [7].

Fig. 2.1 Representation of a
triangular fuzzy number

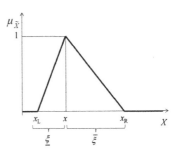

Fig. 2.2 Representation of a
trapezoidal fuzzy number

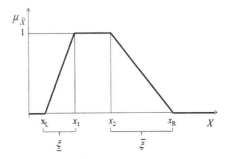

    In particular the values in the interval (x_1, x_2) are considered the core of the fuzzy number, while $\underline{\xi} = x_1 - x_L$ and $\bar{\xi} = x_R - x_2$ are considered the left spread and the right spread, respectively. Note that the values included between x_1 and x_2 belong to $\tilde{X}$ with the highest degree (equal to 1), while the other values lying before x_1 or after x_2 belong to $\tilde{X}$ with a gradually lower degree.

    As Diamond himself suggested, it is possible to extend the distance introduced for triangular fuzzy numbers to the space of trapezoidal ones by the following simple modification:

$$d(\tilde{X}, \tilde{Y})^2 = d\big((x_1, x_2, x_L, x_R)_T, (y_1, y_2, y_L, y_R)_T\big)^2$$
$$= (x_1 - y_1)^2 + (x_2 - y_2)^2 + (x_L - y_L)^2 + (x_R - y_R)^2.$$

    Let's assume now to observe k variables, whose modalities can be expressed as in Fig. 2.2, in a collective of n individuals. Let $\tilde{X}_i = (x_{1i}, x_{2i}, x_{Li}, x_{Ri})_T$ be the k-dimensional fuzzy vector of ith considered unit, where x_{1i}, x_{2i}, x_{Li} and x_{Ri} are k-dimensional vectors of the two cores, the left extremes and the right extremes, respectively.

We propose the following generalization of Diamond's distance:

$$D(\tilde{X}_i, \tilde{X}_j) = \sum_{h=1}^{k} d(\tilde{x}_{ih}, \tilde{x}_{jh})^2 = \sum_{h=1}^{k} d\left((x_{1ih}, x_{2ih}, x_{Lih}, x_{Rih})_T, (x_{1jh}, x_{2jh}, x_{Ljh}, x_{Rjh})_T\right)^2$$

$$= \sum_{h=1}^{k} \left[(x_{1ih} - x_{1jh})^2 + (x_{2ih} - x_{2jh})^2 + (x_{Lih} - x_{Ljh})^2 + (x_{Rih} - x_{Rjh})^2\right], \quad i, j = 1, \ldots, n$$

which still satisfies the required conditions of a distance measure.

2.3   The Use of the Proposed Distance in Hierarchical Clustering

The proposed generalization of Diamond's distance can be used as a measure of dissimilarity between groups in hierarchical clustering. In a previous work [3] we adopted the centroid method; now we implement the Ward's method, which minimizes the total within-cluster variance, by means of *Editor Matlab*.

As the pair of clusters with minimum distance are merged at each step, such a hierarchical procedure finds the pair of clusters that leads to minimum increase in total within-cluster variance. At the initial step, all clusters are singletons (clusters containing a single point), so that the initial distances are defined to be the squared Diamond's distance between points.

It can be shown that the minimization of the increase in the total within-cluster variance occurs by merging at each step the two clusters which present the lowest among the following distances (i.e. the so called merging distance):

$$D(\tilde{X}_i, \tilde{X}_j) = \frac{n_i n_j}{n_i + n_j} \sum_{h=1}^{k} d(\tilde{x}_{ih}, \tilde{x}_{jh})^2. \tag{2.1}$$

The expression (2.1) corresponds to the squared distance between the centroids $\tilde{X}_i$ and $\tilde{X}_j$ of the ith and in the jth group, respectively, multiplied by a quantity which is a function of the numbers n_i and n_j of units of the two groups themselves.

The optimal number g of groups, in which the collective should be divided, is detected in correspondence of the transition from g to $g-1$ groups, reflecting the maximum relative increase:

$$\delta_g = \frac{D_{g-1} - D_g}{D_g}, \quad g = 2, \ldots, n$$

between merging distances. In this case, in fact, the two candidates among g groups are too heterogeneous with each other to be aggregated.

In order to evaluate the goodness of the obtained partition, we compute a fuzzy version of R^2 index. Such a fuzzy clustering index can be still obtained subtracting from one the ratio between the within sum of squares and the total sum of squares, clearly expressed in accordance with the introduced metrics:

$$
\text{FCI} = 1 - \frac{\sum\limits_{j=1}^{g} \sum\limits_{i=1}^{n_j} \sum\limits_{h=1}^{k} d(\tilde{x}_{ijh} - \bar{x}_{jh})^2}{\sum\limits_{i=1}^{n} \sum\limits_{h=1}^{k} d(\tilde{x}_{ih} - \bar{x}_h)^2},
$$

where $\bar{x}_{jh}$ represent the average of the hth variable in the jth group, while $\bar{x}_{jh}$ is the total average of the hth variable. This index provides a measure of variability explained by the obtained partition: the closer it is to one, the better the model fits the observed data.

The just described clustering procedure has been developed through the *MatLab Editor*. In particular, on the basis of the matrices of the centers x_1 and x_2, the left extremes x_L and the right extremes x_R as input parameters, the implemented function provides an output matrix, wherein the first and the second elements of the jth row show the groups (or singletons) aggregated at the jth iteration, while the next three elements show the merging distance, the number of groups identified until the jth iteration and the relative increase between consecutive merging distances, respectively.

At each iteration the minimization of the expression (2.1) allows us to find the lowest within sum of squares arising from every possible combination of two groups among all:

```
it=it+1;
e=[1:nz];
Z(:,1)=e';
Comb=nchoosek(e,2);
nc=size(Comb);
ncc=nc(1,1);
j=1;
dw=0;
while(j<=ncc)
        c=Comb(j,1);
        r=Comb(j,2);
        nc=Z1(c,3);
        nr=Z1(r,3);
        h=1;
        d=0;
        while (h<=k)
                d=d+(((Z1(c,h+3)-Z1(r,h+3))^2)+((Z2(c,h)
-Z2(r,h))^2)
```

```
+((ZL(c,h)-ZL(r,h))^2)+((ZR(c,h)-ZR(r,h))^2));
        h=h+1;
    end;
    dw=[dw,(d)*((nc*nr)/(nc+nr))];
    j=j+1;
end.
```

At this point the two groups which determine the minimum increase in the total within-cluster variance are merged and the new generated group is involved in the form of its centroid in subsequent iterations:

```
dw=dw(:,2:end);
[dmin,imin]=min(dw);
distward=[distward, dmin];
c=Comb(imin,1);
r=Comb(imin,2);
nc=Z(c,3);
nr=Z(r,3);
raggr=[it, Z(c,2),Z(r,2)];
aggr=[aggr; raggr];
xm1=[1, it, (nc+nr)];
xm2=0;
xmL=0;
xmR=0;
j=1;
while (j<=k)
    xm1=[xm1,((nr*Z(r,j+3))+(nc*Z(c,j+3)))/(nr+nc)];
    xm2=[xm2,((nr*Z2(r,j))+(nc*Z2(c,j)))/(nr+nc)];
    xmL=[xmL,((nr*ZL(r,j))+(nc*ZL(c,j)))/(nr+nc)];
    xmR=[xmR,((nr*ZR(r,j))+(nc*ZR(c,j)))/(nr+nc)];
    j=j+1;
end;
Z1=[Z1(1:c-1,:); xm1; Z1(c+1:r-1,:); Z1(r+1:end,:)];
Z2=[Z2(1:c-1,:); xm2(:,2:end); Z2(c+1:r-1,:); Z2(r+1:end,:)];
ZL=[ZL(1:c-1,:); xmL(:,2:end); ZL(c+1:r-1,:); ZL(r+1:end,:)];
ZR=[ZR(1:c-1,:); xmR(:,2:end); ZR(c+1:r-1,:); ZR(r+1:end,:)];
```

Therefore the implemented hierarchical procedure generates all possible partitions of the collective from *n* to 2 groups.

2.3.1 An Application Case to the Italian University System

The university system can fully respond to its responsibility of promoting development only by establishing a connection with manufacturing companies and other institutions of the territory in which it operates. In this perspective, the growth of human capital cannot be reduced to mere relationship between teacher and student established in a classroom, but needs new educational models, alternative to the pure transfer of theoretical knowledge.

In an effort to encourage the organization of learning communities, university regulations allow their students to gain a training experience directly in operational structures. The aim pursued by such regulations is to engage students to adapt in various contexts and to solve operational problems that will occur more frequently during their working life; moreover, the direct experimentation of the theoretical skills stimulates the academic system to recognize the practical needs of the productive system, encouraging the development of new skills.

The effectiveness of internship activities actually organized can be measured not only on the basis of objective data, such as by monitoring how many graduates have pursued the goal of job placement as a result of the gained experience, but also on the basis of subjective data, such as by analyzing the assessments of satisfaction expressed by each of them in relation to that experience.

In this work we analyze the judgments provided by students of Italian universities on the quality of internship activities.

The opinions on internship are collected by the *AlmaLaurea* consortium in a sample survey specifically conducted between 2 and 23 April 2008 (i.e., between 16 and 28 months after graduation). The reference population is represented by 61,347 graduates in calendar year 2006 who said they supported an internship approved by the university, of which 58,904 were potentially reached by e-mail.

The survey, carried out by subjecting 58,904 respondents to a questionnaire prepared for this purpose, is concluded with a response rate of 42.8 %. Note that the valid interviews were weighted by the consortium according to specific characteristics of the reference population correlated with the studied phenomenon, such as the course of study, the university and the faculty of enrollment, gender and so on, in order to avoid as far as possible that the sample is distorted; in particular the answers provided by every graduate interviewed were multiplied by a weighting factor equal to the ratio between the theoretical proportion of the joint distribution (note) of the aforementioned features found in the reference population and the proportion found in the corresponding category.

Respondents express their opinion on a ordinal scale formed by the following categories: "unquestionably unsatisfied," "more unsatisfied than satisfied," "more satisfied than unsatisfied," "unquestionably satisfied." As both Chiandotto and Gola [8] and Lalla et al. [10] proposed a quantification of such an ordinal scale, assigning, respectively, 2, 5, 7, and 10 to the above-mentioned categories, we propose to use the trapezoidal numbers, whose membership function is represented in Fig. 2.3: in particular "unquestionably unsatisfied" is treated as the trapezoidal

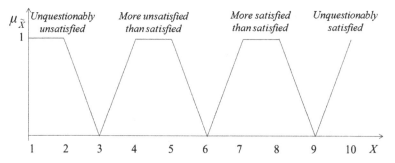

Fig. 2.3 Fuzzy numbers used to express the four response categories

fuzzy number "between 1 and 2," "more unsatisfied than satisfied" as "between 4 and 5," "more satisfied than unsatisfied" as "between 7 and 8" and "unquestionably satisfied" as the triangular fuzzy number "about 10."

In so doing, rather than establishing an unrealistic one-to-one correspondence between verbal terms and numerical values, we associate with each verbal term an interval represented by a neighborhood of the assigned value, the amplitude of which varies depending on the intensity of the expressed judgment. Note that the proposed association of fuzzy numbers is shifted to the right in the interval (1, 10), according to the natural tendency of respondents to medium-high ratings, and also that the values assigned to the two central categories are very close.

2.3.2 Homogeneous Groups of Universities According to the Quality of Internship Activities

At this stage of the work we aim to identify, for each disciplinary area of teaching, homogenous groups of universities according to the opinions (themselves fuzzy) expressed by interviewed graduates on the following aspects concerning internship activities: organization, guidance activities, clearness of formative goals, tutor's helpfulness and competence, autonomy in carrying out the assigned tasks, utility for personal training, prestige of the company and opportunity to convey something useful to colleagues of the company. In particular only those aspects on which the expressed opinions are less correlated with each other (Table 2.1) are taken into account, so that the cluster analysis can be conducted more appropriately: guidance activities, tutor's competence, utility for professional training, prestige of the company, autonomy in carrying out the assigned tasks and opportunity to convey something useful to colleagues of the company.

A university with less than 15 respondents by single disciplinary area is discarded as it is deemed unrepresentative; if instead the respondents are over 15, their answers can be summarized by an average fuzzy trapezoidal vector (which is more informative than the simple centroid of the collected data).

Table 2.1 Correlation between opinions on different aspects

	Organization	Clearness of formative goals	Tutor's helpfulness	Tutor's competence	Utility for personal training	Utility for professional training	Autonomy	Prestige of the company	Opportunity to convey something useful to colleagues
Guidance activities	0.31	0.27	0.20	0.20	0.27	0.30	0.19	0.11	0.18
Organization	–	0.78	0.55	0.53	0.50	0.51	0.43	0.29	0.23
Clearness of formative goals		–	0.56	0.55	0.49	0.50	0.49	0.31	0.23
Tutor's helpfulness			–	0.78	0.41	0.39	0.38	0.28	0.19
Tutor's competence				–	0.42	0.41	0.36	0.25	0.13
Utility for personal training					–	0.75	0.44	0.26	0.29
Utility for professional training						–	0.43	0.22	0.31
Autonomy							–	0.35	0.31
Prestige of the company								–	0.10

Table 2.2 Composition of the obtained groups, by single disciplinary area of teaching

Economics and statistics		Geo-biology	
Less satisfied respondents	More satisfied respondents	Less satisfied respondents	More satisfied respondents
Bari	Bologna	Bari	Calabria
Cagliari	Ferrara	Bologna	Camerino
Cassino	Modena e Reggio Emilia	Cagliari	Lecce
Catania	Padova	Catania	Messina
Firenze	Siena	Ferrara	Modena e Reggio Emilia
Messina	Torino	Firenze	Parma
Perugia	Venezia Cà Foscari	Genova	Perugia
Roma La Sapienza	Verona	Padova	Roma La Sapienza
Trieste	Trento	Sassari	Trieste
Udine	Piemonte Orientale	Siena	Venezia Cà Foscari
Roma Tre		Torino	Catanzaro
		Udine	Piemonte Orientale
		Basilicata	
FCI = 0.72		FCI = 0.55	

Engineering		Literary sciences	
Less satisfied respondents	More satisfied respondents	Less satisfied respondents	More satisfied respondents
Bologna	Calabria	Bari	Calabria
Cagliari	Cassino	Bologna	Catania
Catania	Messina	Cagliari	Firenze
Ferrara	Modena e Reggio Emilia	Ferrara	Parma
Firenze	Salerno	Genova	Siena
Genova	Polytechnic of Torino	Lecce	Torino
Lecce	Reggio Calabria	Padova	Udine
Padova	Roma Tre	Perugia	Viterbo Tuscia
Perugia		Roma La Sapienza	Trento
Roma La Sapienza		Salerno	
Siena		Venezia Cà Foscari	
Trieste		Roma Tre	
Udine			
Trento			
FCI = 0.50		FCI = 0.57	

Medicine		Political and social sciences	
Less satisfied respondents	More satisfied respondents	Less satisfied respondents	More satisfied respondents
Bari	Ferrara	Bari	Bologna Siena
Bologna	Firenze	Catania	Cagliari Torino
Catania	Modena e Reggio Emilia	Ferrara	Calabria Trieste
Genova	Padova	Firenze	Cassino Genova
Messina	Perugia	Padova	Lecce Basilicata
Parma	Siena	Roma La Sapienza	Messina Verona

(continued)

Table 2.2 (continued)

Medicine		Political and social sciences	
Less satisfied respondents	More satisfied respondents	Less satisfied respondents	More satisfied respondents
Roma La Sapienza	Torino	Molise	Salerno Sassari
Chieti e Pescara	Trieste	Roma Tre	Trento Udine
Catanzaro	Udine		Chieti e Pescara
	Verona		Perugia
	Foggia		Venezia Cà Foscari
			Modena e Reggio Emilia
			Piemonte Orientale
			Milano IULM
			Roma LUMSA
			Viterbo Tuscia
FCI = 0.71		FCI = 0.51	

Our attention is specifically directed to the most common disciplinary areas in the Italian academic system: economics and statistics, geo-biology, engineering, literary sciences, medicine, political, and social sciences. For each of the latter we obtain a bipartition of the collective (universities with more or less satisfied respondents) characterized by an FCI always greater than 0.5; in particular such an index exceeds 0.7 in the case of economics and statistics and also of medicine, maybe because the correspondent graduates expressed their opinions in a more discordant way than usual (Table 2.2).

Respondents are generally more satisfied in the universities located in Central and Northern Italy, with a few notable exception: the internship activities organized by the University of Calabria seem to be appreciated whatever the disciplinary area of teaching; conversely the activities supported by the University of Roma La Sapienza are considered not very satisfactory, except those relating to the geo-biological area.

Focusing on engineering, respondents express a greater satisfaction coming from small universities than from big ones (except for the Polytechnic of Torino). This probably happens because the latter have higher expectations and are more critical.

Table 2.3 shows that the final bipartition of universities for each disciplinary area is determined on the basis of the maximum relative increase in the distance between groups, formed throughout the entire agglomeration procedure.

The comparison between trapezoidal fuzzy vectors can be complicated with the naked eye, due to the composite nature of the correspondent components. For this purpose it is appropriate to identify a synthetic value of every component by means of a process called defuzzification.

Among the available methods in the statistical literature, we select the graded mean integration representation [4] for its simplicity, accuracy, and adaptability to

Table 2.3 Distance between groups formed in the last steps of the agglomeration procedure, by disciplinary area of teaching

	Iteration	Number of groups (g)	D_g	δ_g
Economics and statistics	...	...	...	...
	15	6	2.72	0.24
	16	5	3.38	0.64
	17	4	5.56	1.01
	18	3	11.15	0.42
	19	2	15.82	3.46
	20	1	70.59	–
Engineering	...	...	...	...
	16	6	3.59	0.55
	17	5	5.55	0.07
	18	4	5.96	1.88
	19	3	11.21	0.38
	20	2	15.49	2.27
	21	1	50.65	–
Geo-biology	...	...	...	...
	19	6	4.54	0.82
	20	5	8.27	0.07
	21	4	8.87	0.16
	22	3	10.30	0.18
	23	2	12.17	1.50
	24	1	30.39	–
Literary sciences	...	...	...	...
	15	6	3.62	0.01
	16	5	3.66	1.57
	17	4	9.41	0.15
	18	3	10.79	0.30
	19	2	13.98	2.18
	20	1	44.46	–
Medicine	...	...	...	...
	14	6	5.49	0.26
	15	5	6.90	0.02
	16	4	6.75	0.06
	17	3	7.15	1.60
	18	2	18.59	2.15
	19	1	58.63	–
Political and social sciences	...	...	...	...
	26	6	7.90	0.58
	27	5	12.45	0.21
	28	4	15.01	0.20
	29	3	18.05	0.70
	30	2	30.61	1.16
	31	1	66.25	–

Table 2.4 Defuzzificated average vectors for each of the two final groups, by disciplinary area of teaching

		Guidance activities	Tutor's competence	Utility for professional training	Autonomy	Prestige of the company	Opportunity to convey something useful to colleagues
Economics and Statistics	Less satisfied respondents	5.1	6.1	5.1	5.4	8.3	5.7
	More satisfied respondents	7.3	8.4	7.4	8.2	8.4	5.7
Engineering	Less satisfied respondents	6.1	6.4	6.6	7.0	8.0	5.5
	More satisfied respondents	7.8	8.1	7.7	8.1	8.2	6.0
Geo-biology	Less satisfied respondents	5.2	6.7	7.1	6.6	8.1	5.1
	More satisfied respondents	6.7	8.6	8.0	7.7	8.2	5.6
Literary sciences	Less satisfied respondents	5.6	6.2	6.4	6.2	8.3	5.4
	More satisfied respondents	7.5	7.8	7.5	7.2	8.3	6.0
Medicine	Less satisfied respondents	6.1	5.1	5.8	6.7	7.5	6.0
	More satisfied respondents	7.9	8.2	8.3	8.2	7.8	6.1
Political and social sciences	Less satisfied respondents	5.2	5.8	6.1	6.5	8.3	5.6
	More satisfied respondents	6.4	7.4	7.6	7.6	8.4	5.9

trapezoidal fuzzy numbers [5, 6]. In particular the GMIR of a number $\tilde{X}$ coincides with the following form of an expected value:

$$E(\tilde{X}) = \frac{x_L + 2x_1 + 2x_2 + x_R}{6} .$$

We can see it as a weighted mean value, where central components have double the weight of the exterior ones.

Table 2.4 shows trapezoidal defuzzificated average vectors in the two final groups of each disciplinary area.

Whatever the disciplinary area of teaching, the main difference between one group and the other occurs in the satisfaction both with the tutor's competence and with the guidance activities to internship (availability of information, help in the choice).

Conversely the satisfaction both with the prestige of the company and with the opportunity to convey something useful to colleagues seems not to contribute to the identification of the two groups: more specifically respondents' opinion remains in any case high on the first aspect and low on the second one. The last circumstance confirms the difficulties that trainees meet to get involved during the internship experience.

The bipartition of universities is more pronounced relatively to medicine and to economics and statistics rather than to other disciplinary areas, due to the discordance in respondents' opinions on utility for professional training (and also on autonomy in the latter case).

2.4 Conclusions

In this work, looking for an appropriate solution to the problem of partitioning a collective on which features with uncertain borders are observed, we propose a variant of Ward's method.

The operational aspects of such a clustering procedure, founded on a generalization of Diamond's distance to trapezoidal fuzzy vectors and developed through the *MatLab Editor*, are examined within a classification of Italian universities which we carry out on the basis of graduates' judgments (collected by the *AlmaLaurea* consortium in a sample survey between 2 and 23 April 2008) on internship activities.

Respondents express their opinion on a ordinal scale formed by categories, to each of which we associate an interval of values rather than establish an unrealistic one-to-one correspondence between verbal terms and exact numerical values.

Whatever the disciplinary area of teaching taken into account (economics and statistics, geo-biology, engineering, literary sciences, medicine, political and social sciences), we obtain a bipartition of universities as a result of the analysis; such a bipartition is characterized by a fuzzy version of R^2 index always greater than 0.5 and even higher in the case of medicine and economics and statistics, maybe due to a more marked discrepancy in the correspondent graduates' opinions. In overall terms it can be stated that respondents are more satisfied in the universities located in Northern and Central than in Southern Italy, with a few notable exception.

The average profile of each of the two final groups by disciplinary area is expressed by a trapezoidal fuzzy vector that might be defuzzificated at this stage of the analysis (for example through the so-called graded mean integration representation) in order to facilitate comparisons between one and the other. The main difference in such comparisons generally occurs in the satisfaction both with the tutor's competence and with the guidance activities to internship; conversely, the satisfaction both with the prestige of the company and with the opportunity to convey something useful to colleagues seems not to contribute to the identification of the two final groups.

Evidently the obtained results depend on the adopted distance between observation units; in the future we will experiment with other distances as well as with other agglomerative methods (respect to the Ward one), in order to refine the achieved partition.

References

1. Campobasso, F., Fanizzi, A.: A stepwise procedure to select variables in a fuzzy least square regression model. In: Proceedings of the International Conference on Fuzzy Computation Theory and Applications, Paris, France, pp. 417–426 (2011)
2. Campobasso, F., Fanizzi, A., Perchinunno, P.: Homogenous urban poverty clusters within the city of Bari. In: Gervasi, O., Murgante, B., Laganà, A., Taniar, D., Mun, Y., Gavrilova, M.L. (eds.) Computational Science and Its Applications – ICCSA 2008, Part I. LNCS, vol. 5072, pp. 232–244. Springer, Heidelberg (2008)
3. Campobasso, F., Fanizzi, A., Taratini, M.: A proposal for a fuzzy approach to clustering techniques. In: Classification and Data Analysis, Book of Short Papers, University of Padova, Italy, pp. 449–452. Cleup, Padova (2009)
4. Chen, S.H., Hsieh, C.H.: Graded mean integration representation of generalized fuzzy numbers. J. Chin. Fuzzy Syst. Assoc. Taiwan 5(2), 1–7 (1999)
5. Chen, S.H., Hsieh, C.H.: Graded mean representation of generalized fuzzy numbers. In: Proceeding of Sixth Conference on Fuzzy Theory and its Applications (CD ROM), Chinese Fuzzy Systems Association, Taiwan, pp. 1–6 (1998). Filename 031.wdl
6. Chen, S.H., Hsieh, C.H.: Ranking generalized fuzzy number with graded mean integration representation. In: Proceeding of Eighth International Fuzzy Systems Association World Congress (IFSA'99 Taiwan), Taiwan, pp. 551–555 (1999)
7. Chen, S., Wang, C.: Fuzzy distance of trapezoidal fuzzy numbers and application. Int. J. Innovative Comput. Inf. Control ICIC Int. 4(6), 1445–1454 (2008)
8. Chiandotto, B., Gola, M.: *Questionario di base da utilizzare per l' attuazione di un programma per la valutazione della didattica da parte degli studenti*, Italian Ministry of University and Research, Observatory for the evaluation of the Academic System (1999)
9. Diamond, P.M.: Fuzzy least square. Inf. Sci. 46, 141–157 (1988)

10. Lalla, M., Ferrari, D., Pirotti, T.: A fuzzy inference system for teaching evaluation. In: Proceedings of the International Conference on Statistical Modelling for University Evaluation: An International Overview, University of Foggia, pp. 27–30 (2008)
11. Montrone, S., Perchinunno, P., Di Giuro, A., Torre, C.M., Rotondo, F.: Identification of hot spot of social and housing difficulty in urban areas. In: ICCSA 2011. LNCS, vol. 176, pp. 57–78. Springer, Heidelberg (2011)
12. Zadeh, L.A.: Fuzzy sets. Inf. Control **8**(3), 338–353 (1965)

Chapter 3
Geostatistics and the Role of Variogram in Time Series Analysis: A Critical Review

Sandra De Iaco, Monica Palma, and Donato Posa

Abstract Exploratory data analysis and prediction in time series modeling are not typically based on geostatistical techniques, although in several cases applying these techniques might be convenient.

This paper aims to illustrate the usefulness of using Geostatistics and its basic tool, such as the variogram, in time series, especially when an explicit model for the process is not an important goal of the analysis. Moreover, the main differences between the time-domain approach and Geostatistics are highlighted throughout the paper. In order to underline the role of the variogram for modeling and prediction purposes, several theoretical aspects, such as interpolation of missing values, temporal prediction, nonparametric estimation, and their computational problems, are faced through an extensive case study regarding an environmental time series. A modified version of *GSLib* routine for kriging is suitably developed in order to define appropriate temporal search neighborhoods for missing values treatment and prediction.

Keywords *ARMA* model • Kriging • Linear predictors • Nonparametric estimation • Time series • Variogram

3.1 Introduction

In order to describe a stochastic process, one might choose among a simple stochastic representation, a spectral representation or a closed form of the covariance function or variogram. In the case where a simple stochastic representation is not available, it is convenient to deal with a continuous covariance function or even

S. De Iaco (✉) • M. Palma • D. Posa
Department of Management and Economics, University of Salento, Complesso Ecotekne, Via per Monteroni, 73100 Lecce, Italy
e-mail: sandra.deiaco@unisalento.it; monica.palma@unisalento.it; donato.posa@unisalento.it

S. Montrone and P. Perchinunno (eds.), *Statistical Methods for Spatial Planning and Monitoring*, Contributions to Statistics, DOI 10.1007/978-88-470-2751-0_3, © Springer-Verlag Italia 2013

better with a variogram. Linear Geostatistics—variography and kriging—is clearly linked to time series analysis, especially to the time-domain approach where Box–Jenkins methodology [5] is commonly applied.

According to this last approach, it is conventional to inspect the sample autocorrelation function (ACF) and the partial autocorrelation function (PACF), and to infer a model for the process under study. It is well-known that Box–Jenkins techniques for time series analysis essentially use the ACF and PACF for modeling and prediction purposes. In contrast to the use of the ACF and PACF, in Geostatistics the explicit modeling stage is omitted and the variogram can be considered a basic tool to face a variety of inferential problems [7, 26, 37]. The lack of a modeling stage is due to the types of problems Geostatistics was designed to solve: in mining [31], for example, the value of the ore grade is more interesting than the parametric model of the process that produces it. On the other hand, Box–Jenkins analysis was designed to model economic and industrial processes, for which understanding and controlling the process are relevant: this is why, in the time-domain approach, an explicit model is an important goal of the analysis.

Unlike Geostatistics, time series analysis has paid little attention to the variogram; moreover, most of the well-documented and commonly used software on time series analysis does not provide variogram-based predictors in order to face and solve computational aspects related to inferential problems.

However, in the literature, the use of the variogram in time series analysis has been clarified in different ways: Cressie [9] furnishes a graphical procedure for determining nonstationarity in time series analysis; Haslett [20] illustrates the use of the variogram in a time series context and extends the options for its estimation; Ma [34] uses the variogram in construction of stationary time series models, in particular, the author introduces a class of stationary covariance functions, derived from the intrinsically stationary variogram; this class is flexible enough to explain both the short- and the long-term correlation structure of a time series; Ma [35] characterizes a stochastic process having orthogonal increments on the real line in terms of its variogram or its construction; Khachatryan and Bisgaard [28] discuss the variogram as a graphical tool for assessing stationarity in time series analysis. Moreover, the same authors [3] derive a general expression for asymptotic confidence intervals for variogram based on the Delta method for stationary time series. It is also worth citing some recent contributions on strict conditionally negative definiteness and on the use of variogram for spatio-temporal predictions [15, 16, 18].

As it will be highlighted hereafter, different theoretical and practical reasons might justify the use of the variogram even in time series analysis.

The aim of this paper is to enlarge the use of variogram-based geostatistical techniques to analyze time series, not necessarily equally-spaced over time, in order to (a) identify trends and periodicity exhibited by data, (b) describe the regularity of temporal data, (c) estimate missing values, (d) make predictions, (e) estimate the distribution function.

After a brief overview on stochastic processes and stationarity, differences and analogies between time-domain approach and Geostatistics have been presented. Moreover, the role of variogram in time series analysis has been focused and discussed through a case study, where predictions, interpolation of missing values, and nonparametric estimation have been faced. Regarding the computational aspects, a modified version of the *GSLib* routine "KT3D" [13], named "KT3DP," has been used. This last routine has been suitably developed in order to define appropriate temporal search neighborhoods for interpolation and prediction purposes.

3.2 Stochastic Processes

The sequence of observations constituting the time series for statistical analysis may often be considered as a sampling at consecutive (usually, but not necessarily equally-spaced) time points of a much longer sequence of random variables, denoted as X_t, $t \in T$, where T is the time domain. It is frequently convenient to treat this longer sequence as infinite. Such a sequence of random variables is known as stochastic process with a discrete time parameter. A stochastic process of a continuous time parameter t can be defined for $0 \leqslant t < \infty$ or $-\infty < t < \infty$. A sample from such a process could consist of observations at a finite number of times, or it could consist of a continuous observation over an interval of time.

In the literature, there has been a revived interest in continuous-time processes, which have also been utilized very successfully for modeling irregularly-space data [24] or when the physical model takes the form of a stochastic differential equation [21] or with the aim to develop spatio-temporal random fields [36]. Moreover, a stochastic process of a discrete time parameter may often be thought of as a sampling at equally-spaced time points of a stochastic process of a continuous time parameter.

3.2.1 Basic Notions

Let $\{X_t, t \in T\}$ be a real-valued stochastic process over a temporal domain $T \subseteq \mathbb{R}$, with covariance, variogram and ACF defined, respectively, as follows:

$$C(t_1, t_2) = E[(X_{t_1} - E(X_{t_1}))(X_{t_2} - E(X_{t_2}))], \quad t_1, t_2 \in T$$
$$\gamma(t_1, t_2) = 0.5 \operatorname{Var}[X_{t_1} - X_{t_2}], \quad t_1, t_2 \in T$$
$$\rho(t_1, t_2) = \frac{C(t_1, t_2)}{\sqrt{\operatorname{Var}(X_{t_1})\operatorname{Var}(X_{t_2})}}, \quad t_1, t_2 \in T.$$

Techniques, collectively known as time series analysis, have their foundations in the theory of stationary processes [57]. In particular, a stochastic process $\{X_t, t \in T\}$ is strict stationary if the finite-dimensional distributions are invariant under an arbitrary translation of the time points. Second-order stationarity and intrinsic stationarity are usual work hypothesis of a stochastic process in terms of the covariance function (or ACF) and the variogram, respectively. The stochastic process $\{X_t, t \in T\}$ is second-order stationary if its covariance $C(t_1, t_2)$ depends solely on the temporal lag $(t_1 - t_2)$ and its expected value is constant. Similarly, the stochastic process $\{X_t, t \in T\}$ is intrinsically stationary if its variogram $\gamma(t_1, t_2)$ depends solely on the temporal lag $(t_1 - t_2)$ and the expected value of the difference $(X_{t_1} - X_{t_2})$ is constant.

Historically, the study of second-order stationary processes originated in the study of Gaussian processes, for which second-order stationarity entails strict stationarity. This is not surprising, considering how much of applied probability and statistics has its origins in the study of Gaussian random variables. Nevertheless, the stationarity hypothesis is not without any theoretical justification: many processes—positive recurrent Markov chains and certain diffusion processes among them—exhibit limiting stationary behavior.

3.2.2 Alternatives in Stochastic Representation

A stationary process $\{X_t, t \in T\}$ might be described through a simple stochastic representation (as for the stationary discrete-time autoregressive and moving average time series—Sect. 3.3.2), a spectral representation or a closed form of the covariance function or variogram.

In the discrete case, the well-known general linear model of a process X, with zero expected value, is given as

$$X_t = \sum_{i=1}^{\infty} \theta_i Z_{t-1}, \tag{3.1}$$

where Z is a purely random process (often called white noise); by analogy, one could define the general linear model in continuous time, that is

$$X_t = \int_0^{\infty} h(u) Z(t - u) \, du, \tag{3.2}$$

where Z is a continuous white noise and $h(\cdot)$ is an absolutely integrable weight function. Although it turns out to be a useful mathematical construction for some theoretical and practical purposes, there has been a historical debate on whether it is possible to think of a continuous white noise, whose ACF is discontinuous (1 in zero and 0 otherwise), and whose variance is not finite.

Moreover, by recalling the spectral representation theorem, a stationary process X can be written in the form of the following Fourier–Stieltjes integral:

$$X_t = \int_{-\infty}^{\infty} e^{i\lambda t} Z \, (du), \tag{3.3}$$

where Z is an orthogonal random spectral measure and i is the unit pure imaginary number.

If a simple representation of the stochastic process is not available or it cannot be derived from physical laws or previous experiences, dealing with a continuous covariance function or even better with a variogram is convenient. In the literature, there are wide classes of covariance functions and variograms [29, 50, 56], which are characterized by different properties in terms of behavior at the origin or to infinity.

As it will be clarified hereafter, estimating and modeling the variogram are crucial steps of traditional structural analysis, since the variogram model might be used for prediction purposes. Similarly to spatial analysis, the variogram model might then be used for temporal prediction which could be performed through several forms of interpolators [38, 41, 42]. Indeed, there are two developments that lead to the same functional form: radial basis functions and the regression method known as kriging. The key of the interrelationship lies in the positive definiteness of the kernel function [43].

3.3 Time Series Analysis Versus Geostatistics

Both time series analysis and Geostatistics provide some tools and techniques for analyzing second-order stationary stochastic processes. Hence, it is interesting to propose a comparison between them.

Time series analysis may be approached through the frequency domain, in which case it is called harmonic analysis [4], or spectral analysis [19, 23, 48] or through the time-domain, in which case the Box–Jenkins methodology is commonly applied [5]. On the other hand, Geostatistics may be approached through the structural analysis often used in multidimensional space [7, 26, 37].

It is shareable that time-domain approach has some elements in common with Geostatistics. For this reason, some details on similarities, as well as dissimilarities, between linear Geostatistics—variography and kriging—and the time-domain approach of time series analysis are presented in the following.

3.3.1 General Similarities and Dissimilarities

In time-domain approach and according to Box–Jenkins techniques for time series analysis, it is conventional to inspect the ACF and the PACF, and to infer a model for the process under study. In other words, the ACF and PACF are used for modeling and prediction purposes. Similarly, in Geostatistics, prediction is directly based on the correlation structure or variogram; however, the explicit modeling stage is commonly omitted.

The greatest difference between Geostatistics and the time-domain approach is that this last approach works essentially in one dimension and it is based on a fundamental property of time, that is the temporal order; on the other hand,

Geostatistics applies to multiple dimensions. This is convenient when the data have a spatio-temporal structure; in this case Geostatistics offers a theoretical basis, not only for a temporal study, but for a more general space–time analysis [8, 12, 30]. Moreover, in a multiple dimension representation, time series values can be interpreted as a finite realization of a random field X anchored to different time units (hour, day, week, month, year). For example, an hourly time series available for a year might be viewed as a realization of $X(h, w)$, where $h = 1, 2, \ldots, 168$ identifies the hour within the week and $w = 1, 2, \ldots, 52$ the week within the year, or as a realization of $X(h, w, d)$, where $h = 1, 2, \ldots, 24$ identifies the hour within the day, $w = 1, 2, \ldots, 52$ the week within the year and $d = 1, 2, \ldots, 7$ the day of the week. Such a representation on a two- or three-dimensional domain might be useful for missing values analysis or to study anisotropies and/or periodicities along different patterns.

In addition, time series approach is usually based on the analysis of regularly spaced data, where the regular spacing of the data corresponds sometimes to the unit time interval. The matter of data spacing is concerned with the distinction between discrete and continuous processes. Although, within the Box–Jenkins scheme, it is possible to model a temporal process as continuous by using an appropriate linear differential equation, Geostatistics avoids this problem by treating any process as continuous and modeling the variogram as a continuous function.

3.3.2 Approaches in Problem Solving

Both time-domain approach and geostatistical techniques face the same classes of problems, such as exploratory data analysis, modeling and prediction [51].

Exploratory data analysis is often useful to analyze a process in detail and decide which kind of approach is more appropriate, or to solve problems of classification and recognition. In Geostatistics, the variogram provides such a characteristic, while in Box–Jenkins analysis the ACF and the PACF play this role.

In *modeling*, the goals include understanding the nature of the process, controlling the process by sequential decisions, filtering the process to rid it of noise and predicting unknown values of the process. In this case, an explicit parametric expression of the process is required: the form of the model is chosen on the basis of some structural characteristics of the process among a family of models and the model parameters are estimated by using appropriate fitting procedures. Box–Jenkins analysis is a modeling procedure, based on the family of stationary linear models composed of:

- the *moving average* (MA) process of order q,

$$X_t = Z_t - \theta_1 Z_{t-1} - \cdots - \theta_q Z_{t-q},$$

that is $X_t = \theta(B)Z_t$, where B is the backshift operator $(BX_t = X_{t-1})$, $\theta(B)$ is the polynomial $1 - \theta_1 B - \cdots - \theta_q B^q$, θ_i, for $i = 1, 2, \ldots, q$, are parameters such that the invertibility of the process is guaranteed, i.e. the roots of the corresponding characteristic equation, $\theta(B) = 0$, lie outside the unit circle;

- the *autoregressive* (AR) process of order p,

$$X_t = \phi_1 X_{t-1} + \cdots + \phi_p X_{t-p} + Z_t,$$

that is $\phi(B)X_t = Z_t$, where B is the backshift operator, $\phi(B)$ is the polynomial $1 + \phi_1 B + \cdots + \phi_p B^p$, ϕ_i, $i = 1, 2, \ldots, p$, are parameters such that the stationarity of the process is guaranteed, or equivalently the roots of the corresponding characteristic equation, $\phi(B) = 0$, lie outside the unit circle;

- the *autoregressive-moving average* (ARMA) process, which is a mixture of the two above-mentioned processes,

$$X_t - \phi_1 X_{t-1} - \cdots - \phi_p X_{t-p} = Z_t - \theta_1 Z_{t-1} - \cdots - \theta_q Z_{t-q},$$

that is $\phi(B)X_t = \theta(B)Z_t$, where B is the backshift operator, ϕ_i and θ_j, for $i = 1, 2, \ldots, p$ and $j = 1, 2, \ldots, q$, are parameters such that the invertibility and the stationarity of the process are guaranteed, i.e. the roots of the corresponding characteristic equations, $\phi(B) = 0$ and $\theta(B) = 0$, lie outside the unit circle.

Note that an ARMA model involves fewer parameters than a MA or AR process itself, since an MA process of finite order can be expressed as an AR process of infinite order, while an AR process of finite order can be expressed as an MA process of infinite order. Both MA and AR processes are special cases of the general linear process (3.1). However, it is well-known that a continuous ARMA process satisfies an appropriate linear differential equation; hence, its application requires the knowledge of the underlying physical, dynamic model which describes the temporal evolution of the variable of interest.

As already pointed out, the modeling stage is not explicit in Geostatistics; however, the dual form of kriging can be used. In this case, the general form of the radial basis function [42, 43] depends solely on the kernel function and does not require a model for the process. The general form of the radial basis function interpolator is the following:

$$\widehat{X}(t) = \sum_{i=1}^{n} b_i g(t - t_i) + \sum_{k=0}^{p} a_k f_k(t),$$

where the function g must satisfy the positive definiteness condition, X_t is the function to be interpolated and the t_i, $i = 1, 2, \ldots, n$, are the data time points. The $f_k(t)$, $k = 0, 1, \ldots, p$, are linearly independent functions, usually taken to be monomials in t. Micchelli [40] has shown that the coefficients b_i, $i = 1, 2, \ldots, n$, and a_k, $k = 0, 1, \ldots, p$, are determined, given the conditions on g and f_k.

In *prediction*, the objective is to estimate the unknown value x_t, at time t, of the stochastic process X, using the data observed in the past (extrapolation mode) and, in case of interpolation, the data observed after the time point t. This can be achieved either using Box–Jenkins analysis or, even without any explicit modeling of the process, using linear Geostatistics, where only the knowledge of the variogram model is required. Indeed, given the linear predictor $\widehat{X}_t$ of the intrinsic stationary process X at the time point t:

$$\widehat{X}_t = \sum_{i=1}^{n} \lambda_i(t) X_{t_i}, \tag{3.4}$$

where $\lambda_i(t)$, $i = 1, 2, \ldots, n$, are unknown real coefficients and X_{t_i} are random variables of the process X at the sampled time points t_i (before and after the time point t, in case of interpolation, or related only to the past in case of extrapolation), the unknown coefficients or weights $\lambda_i(t), i = 1, 2, \ldots, n$, of (3.4) are obtained by solving the following kriging system

$$\begin{bmatrix} \gamma_{11} \cdots \gamma_{1n} - 1 \\ \gamma_{21} \cdots \gamma_{2n} - 1 \\ \vdots \quad \ddots \quad \vdots \quad \vdots \\ \gamma_{n1} \cdots \gamma_{nn} - 1 \\ 1 \ \ldots \ 1 \quad 0 \end{bmatrix} \begin{bmatrix} \lambda_1 \\ \lambda_2 \\ \vdots \\ \lambda_n \\ \mu \end{bmatrix} = \begin{bmatrix} \gamma_{10} \\ \gamma_{20} \\ \vdots \\ \gamma_{n0} \\ 1 \end{bmatrix}, \tag{3.5}$$

where $\gamma_{ij} = 0.5 \operatorname{Var}(X_{t_i} - X_{t_j})$, $\gamma_{i0} = 0.5 \operatorname{Var}(X_{t_i} - X_t)$, μ is the Lagrange multiplier. Note that if the variogram function γ is conditionally strictly negative definite, then the above system presents one and only one solution.

This is what is known in literature as ordinary kriging [26, 37] and it is used when the expected value of the process is constant and unknown (which is the most common case in practice). Indeed, ordinary kriging and kriging with a trend can be viewed as two minimum error variance algorithms which apply the same normal equations, with different constraints, to obtain the kriging weights [26].

Since the kriging system can be expressed in terms of the variogram, as in (3.5), the kriging predictor can be used even when the stochastic process under study satisfies the intrinsic hypothesis. Moreover, using a predictor based on a variogram, rather than on a covariance, avoids the estimation of the expected value, if this last is unknown. Nevertheless, it is relevant to remind that Gevers [17] showed that the predictors based on known variogram or covariance are identical when the unknown mean is replaced by its minimum variance estimator.

In the prediction stage, another difference between Geostatistics and the time-domain approach lies in the form of the estimators of unknown values. The ARMA estimator seems to be more general than the linear kriging estimator (3.4), which has always an autoregressive form. Indeed, if the configuration of the time points in the neighborhood does not change, the kriging weights are fixed; hence, the linear kriging estimator has an autoregressive form. Note that the non-zero kriging weights, associated with the points in the neighborhood, give an idea of the order of the corresponding autoregressive process. In addition, the kriging autoregressive form of X_t is flexible enough to consider, in the modeling stage, variables at time points before and after the time point t.

Moreover, it is well-known that in the multi-Gaussian case, the conditional expectation—which is the best in the classical sense—of the unknown value, given the data, is precisely the kriging estimator. However, since any invertible ARMA process has an AR representation, the limitation of the kriging estimator is not really important.

3.3.3 Nonstationary Stochastic Processes

If the stationarity hypothesis is not reasonable for the time series under study, the integrated mixed models might be a possible modeling choice in Box–Jenkins analysis. These models are called *autoregressive integrated moving average* (ARIMA) processes; more specifically, in presence of a periodic component, they are called *seasonal autoregressive integrated moving average* (SARIMA) processes. To deal with periodicity and trend, Box and Jenkins suggest to difference d times the process, at appropriate lags, in order to yield a stationary process which may then be modelled as an ARMA process. In other words, the dth difference of ARIMA and the dth seasonal difference of SARIMA are stationary mixed ARMA processes. The role of periodic components in a stochastic process is reduced to that of a deterministic component, that is factored out by differencing, so that the ACF and the PACF of the stationary residuals are analyzed. This is justified by the Yule's model, where periodicity is treated as a purely deterministic component and the purely stochastic component is modelled as an MA process. Moreover, this attitude stems from the fact that Box–Jenkins analysis was first developed for the analysis of economic time series, in which periodicity usually reflects fairly regular seasonal influences.

Box–Jenkins analysis of nonstationary stochastic processes modeling is closely related to the intrinsic random function of order k, introduced in Geostatistics by Matheron [39], where the trend component in a process is usually interpreted as a deterministic component, modelled with a local polynomial form. If the expected value of the process is not constant but it can be expressed as a polynomial functional form, the more general kriging with a trend model (or universal kriging) could be applied. In both cases, the kriging predictor is built to be unbiased and with the minimum error variance. Similarly, the variogram of the stationary residual component is used in the universal kriging system. As regards periodicity, this component can be factored out (by using, for example, the moving average method) or alternatively it can be described by a periodic variogram [14]. In [27, 46], it was shown that, in interpolation situations, the overall contribution of the trend model to the estimate at an unsampled point, surrounded by data, is similar for both ordinary kriging and kriging with a polynomial trend; moreover, if the residual variogram model is the same, then the estimates obtained by using the two methods are statistically equivalent, no matter how the trend may appear within the local neighborhood of the point of interest.

3.4 The Use of Variogram: Main Advantages

Most of geostatistical techniques are based on the use of variogram [7, 26, 37] and different reasons might justify the use of the variogram even in time series analysis. First of all, the variogram is usually preferable with respect to the covariance [11], since it can describe a wider class of stochastic processes: the class of intrinsic stochastic processes, for which only the variogram is defined, includes the class of second-order stationary stochastic processes. This is why it was introduced by Kolmogorov [29] for the study of turbulent flow and by Gandin [16] for meteorological applications. The second reason is based on a practical aspect: the variogram, unlike the covariance, does not require the knowledge of the expected value of the associated stochastic process. As a consequence of above, the variogram was called *the structure function* by Yaglom [57].

Some details about the advantages of using the variogram even in time series analysis are given below.

3.4.1 Estimation Aspects

Before choosing an appropriate temporal correlation model, it is essential to estimate the corresponding second-order moments from data.

In the time-domain approach, the sample ACF is commonly a standard exploratory tool for identifying the model structure of the temporal process under study, as well as, in Geostatistics, the sample covariance and variogram are basic tools used essentially for modeling and prediction purposes in the kriging system. However, the use of variogram estimator is much more convenient, as it will be clarified hereafter. First of all, under second-order stationarity, if the expected value is unknown and it is estimated from the data, then this introduces a bias in the covariance estimator [56]; even for long time series (100–200 sample data) this bias can be surprisingly large. On the other hand, the variogram is not affected by this problem, since it automatically filters the expected value and the unbiasedness is guaranteed for its classical estimator [37]:

$$\widehat{\gamma}(r_t) = \frac{1}{2|M(r_t)|} \sum_{M(r_t)} [X_{t+h_t} - X_t]^2, \tag{3.6}$$

where r_t is the temporal lag, $M(r_t) = \{t + h_t \in H$ and $t \in H$ such that $\|r_t - h_t\| < \delta_t\}$, δ_t is the tolerance, H is the set of data at different time points (not necessarily equally-spaced), and $|M(r_t)|$ is the cardinality of this set.

Moreover, in nature most processes are nonstationary; in this case, the use of the ACF, or the covariance, can potentially be misleading, since a nonstationary time series is theoretically characterized by not having constant mean and variance. Although one can obviously always compute the sample ACF or covariance, plotting the sample ACF or covariance, against the temporal lags, for nonstationary time series is inaccurate, since the autocorrelations depend on the supporting time points. When the process is not mean stationary, the classical estimators of ACF or covariance become hopelessly biased, moreover if the process is not variance stationary, the covariance, as well as the ACF, is not even defined. As discussed in [28], if the nonstationary integrated MA process is considered, i.e.

$$X_t = X_{t-1} + Z_t - \theta Z_{t-1}, \quad t = 1, 2, \ldots,$$

where Z_t is a white noise process, then the covariance $C(X_t, X_{t-h})$ is a function of both h and t, since

$$C(X_t, X_{t-h}) = \sigma^2(1 - \theta)[1 + (t - h - 1)(1 - \theta)].$$

In contrast to the ACF and to the covariance, the variogram is not only a complementary exploratory tool, but it is well defined for several nonstationary processes, including ARIMA processes. Indeed, the variogram is well defined for the much wider class of so-called intrinsic processes, its classical estimator is unbiased when the process is only mean stationary and an alternative estimator has only a small bias even when the process is neither mean nor variance stationary. Haslett [20] illustrated the properties of different variogram estimators.

3.4.2 Trends and Periodicity

The role of variogram to identify trends and periodicity exhibited by data is surely not negligible.

Khachatryan [28] consolidates the use of the sample variogram as a graphical tool for assessing stationarity in time series analysis. It is well-known that if the sample variogram $\hat{\gamma}(r_t)$ increases more rapidly than r_t^2, then intrinsic stationarity is not admissible. In this case, analyzing and fitting the trend component that characterizes the process of interest are necessary. After removing the trend component, the sample variogram of the stationary residual component is computed and modelled; then the corresponding variogram model is used for interpolation and/or prediction purposes.

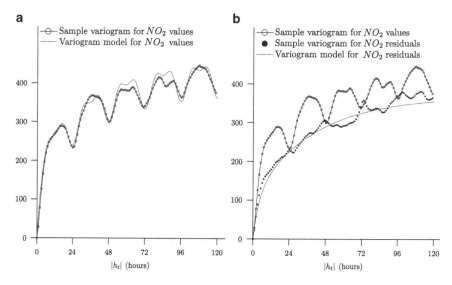

Fig. 3.1 Sample temporal variograms and models. (**a**) Variogram for NO₂ hourly concentrations. (**b**) Variogram for NO₂ residuals

As regards the periodicity, this component can be factored out using, for example, the moving average method [6], or alternatively it can be retained and described by a periodic variogram model. It is worth noting that, in the former case, the variogram of residuals might be viewed as the convolution of the variogram with periodicity. For example, in Fig. 3.1 the sample temporal variograms for NO_2 hourly averages ($\mu g/m^3$) and NO_2 residuals, measured during January 2000 at a monitoring station in the district of Milan, Italy, are illustrated, together with their models. Further details about the analytic form of the variogram models are given in Sect. 3.5.1.

In the case study presented hereafter, the use of periodic and nonperiodic variogram models has been proposed for both interpolation and prediction purposes and a new *GSLib* routine for kriging, named "KT3DP," has been suitably developed in order to define appropriate temporal search neighborhoods in presence of periodicity.

3.4.3 Continuity and Scales of Variation

Unlike Box–Jenkins approach, in Geostatistics the behavior of the variogram function near the origin is analyzed in order to describe the continuity of the variable under study [44]. Moreover, the variogram is a useful tool for assessing the scale of variation which characterizes the time series. Regarding this aspect, it is interesting to recall the following decomposition of the process $\{X_t,\ t \in T \subseteq \mathbb{R}\}$,

$$X_t = \mu_t + Y_t + \eta_t + \varepsilon_t, \tag{3.7}$$

where:

- $\mu_t = E(X_t)$ is the deterministic mean, also called large-scale variation.
- Y is a zero mean intrinsically stationary process, whose variogram range (if it exists) is larger than the minimum temporal lag and it is called small-scale variation.
- η is a zero mean intrinsically stationary process independent of Y, whose variogram range (if it exists) is smaller than the minimum temporal lag and it is called microscale variation.
- ε, called measurement error, is a zero mean white noise process which is independent of Y and η.

Clearly, decomposition (3.7) is not unique and it involves the specific features of the process under study.

If μ_t is assumed constant over the temporal domain, then the following variogram for X is obtained:

$$\gamma_X(h_t) = \gamma_Y(h_t) + \gamma_\eta(h_t) + K_{ME},$$

where $K_{ME} = \text{Var}(\varepsilon_t)$, which means that a nugget effect [26] can be recognized, i.e.

$$\lim_{|h_t| \to 0} \gamma_X(h_t) = K_{ME} \neq 0.$$

However, since estimation and modeling are based on the available data, nothing can be generally said about the variogram at temporal lag smaller than the minimum lag. Thus, assuming that

$$\lim_{|h_t| \to 0} \gamma_\eta(h_t) = K_{MS} \neq 0,$$

the total nugget effect is given by $K_{ME} + K_{MS}$. In these cases, if $\gamma(0)$ is set equal to $K_{ME} + K_{MS}$ in the right-hand side of system (3.5), kriging does not yield exact interpolation, but it can smooth the data which contain errors or an unknown variability [9].

Moreover, the presence of a nugget effect can suggest an improvement in the sampling scheme or corrections of the measurement errors or detections of outliers.

Janis and Robeson [22] determined the nugget effects of the variogram models, associated with different subsets of the temporal interval of interest, and studied the time series of the nugget effects to evaluate the representativeness of historical air-temperature records.

3.4.4 Missing Values Estimation

The estimation of missing data represents surely one of the main issues to be addressed in a variety of areas of time series analysis, ranging from engineering to economics and to environmental sciences. In literature, there is a large number of methods for data reconstruction, either based on classical time series approaches for stationary or nonstationary processes or based on spatial/spatial–temporal interpolation methods, such as simple, ordinary, or universal kriging. A review of classical missing data techniques, such as listwise deletion, mean imputation, regression analysis, and expectation maximization, is given in [32]; other contributions can be found in [33], where the author proposed a Ljung's method to estimate maximum values, in [49], where the author proposed a modification of the singular spectrum analysis for time series with missing data, or in [2, 54, 55].

However, the use of kriging to estimate missing values in time series is convenient for different reasons: availability of the error variance of the estimates, reconstruction of data referred to regular/irregular spacing, estimation of the unknown value at time t, using data observed before and after the time point t, even without any explicit modeling or assumption on the probability distribution of the process.

In this context, estimating and modeling the variogram of the stochastic process are fundamental, since only if the variogram model is appropriate, one can rely on further kriging results. However, the analyst should pay attention to the well-known screening effect [7, 52]. In this case, positive kriging weights (significantly different from zero) are concentrated on a restricted subset of data near to the estimation point and the contribution of the remaining data are screened off. In one-dimensional space, which is the case of time series, if a linear variogram model is used (related to a Brownian motion), the ordinary kriging estimator only depends on the two contiguous realizations; analogous results occur for simple kriging, if an exponential model is considered, because of the Markov property [7].

Moreover, if the time series is interpreted as a finite realization of a random field X anchored to different time units (hour, day, week, month, year), the study of anisotropies along different patterns [47], which often correspond to periodicities in time series, and the use of an anisotropic variogram might be useful in missing values interpolation. For example, an hourly time series, available for a year, might be viewed as a realization of $X(h, w)$, where $h = 1, 2, \ldots, 168$ identifies the hour within the week and $w = 1, 2, \ldots, 52$ the week within the year, as shown in Fig. 3.2. In this case, a variogram model with zonal anisotropy could be fitted and used in the kriging system.

The use of kriging for data values reconstruction will be widely discussed in the case study.

3.4.5   Nonparametric Estimation and Simulation

The usefulness of variogram in time series analysis can be appreciated especially when the aim concerns typical geostatistical techniques, such as nonparametric estimation or simulation of the variable under study.

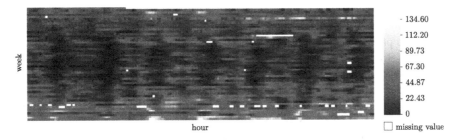

Fig. 3.2 Hourly NO$_2$ averages measured in 2000 (plotted per week)

The kriging approach, based on the knowledge of variogram, leads naturally to nonparametric estimation as well as to different simulation techniques. Indicator kriging is a nonparametric approach to estimate the posterior cumulative distribution function (c.d.f.) of the variable under study at an unsampled point [25, 45]. In this context, given the observed time series x_{t_i}, t_i, $i = 1, 2, \ldots, n$, the conditional probability Prob$\{X_t \leq x | \mathcal{H}_n\}$ with $\mathcal{H}_n = \{x_{t_i}, t_i, i = 1, 2, \ldots, n\}$, is interpreted as the conditional expectation of an indicator random field $I(t; x)$,

$$\text{Prob}\{X_t \leq x | \mathcal{H}_n\} = E[I(t; x | \mathcal{H}_n)],$$

where

$$I(t; x) = \begin{cases} 1, \text{ if } X_t \leq x \\ 0, \text{ if } X_t > x, \end{cases} \tag{3.8}$$

and it is modelled by a linear combination of neighboring indicator data values. The weights of this combination are given by a kriging system, equivalent to (3.5), where the structural function used is the indicator variogram associated with the indicator data values for a given threshold value x:

$$\gamma_I(h_t; x) = 0.5 \, E[I(t + h_t; x) - I(t; x)]^2, \quad \forall h_t \in \mathbb{R}. \tag{3.9}$$

This approach might be useful to support national policies for environmental and health protection, which, for example, have to keep pollution concentrations down the specific thresholds, called *levels of attention*, according to national or international directives. Then, for a given time series of an environmental variable, it might be useful to estimate the probability that the variable under study exceeds a fixed limit, so that appropriate and prompt solutions might be adopted. For example, decisions about traffic limitation in high traffic urban area might be supported by the knowledge of the probability that a hazardous pollutant exceeds the level of attention.

Moreover, variogram-based simulation algorithms, such as the sequential indicator simulation [18] and the LU decomposition algorithm [1], are useful to study the temporal variability without choosing a model for the process. Clearly, the goal of any simulation algorithm is to reproduce the global features of a phenomenon, usually in terms of the first- and second-order moments of the corresponding stochastic process [15, 26]. In time series, conditional simulation might be fundamental, if it is used as a tool to evaluate the impact of temporal uncertainty on the results of complex procedures. For example, if a variogram model for the time series under study can be easily derived, the choice of the variogram-based simulation algorithms in time series is justified to check the variability of the temporal evolution with respect to the expected results.

The case study, proposed hereafter, will discuss the indicator kriging approach and its capability for assessing the probability that the variable of interest exceeds specific threshold values.

3.5   A Case Study

The environmental monitoring network of Lombardy region (Italy) collects data concerning hazardous pollutants and atmospherical variables, in order to continuously control the air quality of urban, suburban, and industrialized areas of the region.

In this paper, nitrogen dioxide (NO_2) hourly concentrations ($\mu g/m^3$) measured at one of the monitoring stations of the district of Milan during January 2000, have been analyzed using geostatistical techniques. In particular, the station "Parco Lambro", located in an urbanized residential zone and close to one of the main urban areas of the district of Milan, has been considered.

The case study aims to point out the potentiality of geostatistical techniques, and especially the role of variogram, to solve estimation and prediction problems in time series analysis, even in presence of a periodic component. Hence,

- structural analysis,
- estimation of some consecutive values assumed as missing,
- prediction of NO_2 hourly averages,
- estimation of the c.d.f. of NO_2 hourly averages at some unsampled time points,

will be discussed.

3.5.1   Structural Analysis

As previously pointed out, the variogram is a more general measure of correlation than the covariance and it is a useful graphical tool for assessing stationarity and periodicity in time series analysis. The latter aspect is immediately confirmed by the structural analysis herein developed for the time series of NO_2 hourly measurements.

The sample temporal variogram, estimated from the data, reflects the diurnal periodicity at 24 h of the variable under study, as illustrated in Fig. 3.1a. Hence, two different alternative approaches have been considered:

1. the periodic component has been factored out,
2. the periodic component has been retained and a variogram function with hole effect has been chosen to model the time series of the original data.

In this last case, the following model has been used:

$$\gamma(h_t) = 127 \, \text{Exp}(|h_t|/6) +$$
$$+ 247 \, \text{Exp}(|h_t|/120) + 18.5 \, \text{Cos}(|h_t|/12) + 38 \, \text{Cos}(|h_t|/24), \qquad (3.10)$$

where $\text{Exp}(\cdot)$ and $\text{Cos}(\cdot)$ are the shortened forms of the exponential and the cosine variogram models [10], respectively. Figure 3.1a shows the sample temporal variograms for NO_2 hourly observations, together with the fitted model.

On the other hand, before performing structural analysis, the time series under study has been previously deseasonalized. The diurnal component which characterizes NO_2 hourly concentrations has been estimated and removed by the *FORTRAN* program "REMOVE" described in [14]. Using this program, the NO_2 time series has been firstly segmented in 4 homogeneous sub-periods (i.e., from the 1st to the 168th hour, from the 169th to the 336th hour, from the 337th to the 504th hour, and from the 505th hour to the last time point of the month under study) and successively the moving average method [6] has been applied to sequences with at least 60 consecutive values within each sub-period. For each of these sequences the diurnal component has been separately computed and removed.

Figure 3.1b shows the sample temporal variograms for NO_2 hourly residuals, together with the fitted model

$$\tilde{\gamma}(h_t) = 108 \, \text{Exp}(|h_t|/12) + 260 \, \text{Exp}(|h_t|/120). \qquad (3.11)$$

Note that in both cases (original data and residuals), the behavior of the variogram functions near the origin is linear with no nugget effect.

In order to evaluate the goodness of models (3.10) and (3.11), cross-validation has been performed and estimates for NO_2 hourly concentrations and NO_2 residuals, respectively, at all data points have been obtained by using the kriging technique. Figure 3.3 shows the scatter plots of NO_2 observed values (a) and NO_2 residuals (b) towards the corresponding estimated values. The high values of the linear correlation coefficients (0.977 and 0.976, respectively) confirm the goodness of the above fitted models.

It is important to point out that the variogram model (3.10) has been validated using a modified version of the *GSLib* program "KT3D" [13], named "KT3DP." This program has been developed in order to properly define the neighborhood, i.e. the subset of available data used in the kriging system. By taking into account the main features of the analyzed pollutant and its temporal behavior (periodicity at 24 h),

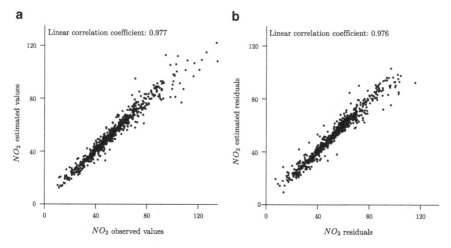

Fig. 3.3 Scatter plots between observed and estimated values. (**a**) Diagram of NO$_2$ hourly concentrations towards the estimated ones. (**b**) Diagram of NO$_2$ residuals towards the estimated ones

the kriging routine has been modified in such a way that, the value at time t is estimated by considering data observed

- at the two adjacent time points, $(t-1)$ and $(t+1)$,
- at the same hour of the day before and/or later, $(t-d)$ and $(t+d)$ with $d = 24$, and some hours before and/or later, $(t-d \pm k)$ and $(t+d \pm k)$ with $k = 1, 2, 3$,
- at the same hour of two days before and/or later, $(t-2d)$ and $(t+2d)$ with $d = 24$, and some hours before and/or later, $(t-2d \pm k)$ and $(t+2d \pm k)$ with $k = 1, 2, 3$,

up to a maximum number of eight values.

On the other hand, variogram model (3.11), which describes the temporal correlation for NO$_2$ residuals, has been validated using the *GSLib* program "KT3D."

3.5.2 Estimation of Missing Values

For several technical reasons, it happens that a monitoring station does not work for an interval of time or records data which might not be considered valid. In this case, especially for environmental variables, it is crucial to reconstruct the time series by estimating the missing values. Among the methods known in literature [32], kriging could be a very useful tool for time series reconstruction.

In order to illustrate the usefulness of kriging as data reconstruction method, various series of consecutive missing values have been analyzed for either the original time series with a 24 h periodic behavior or the deseasonalized time series.

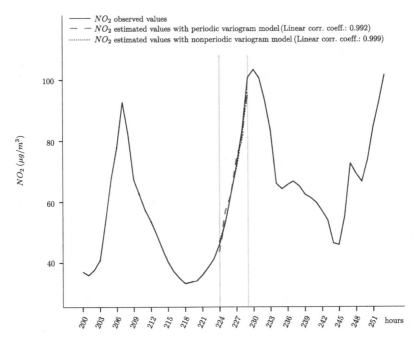

Fig. 3.4 Time plot of NO_2 estimated missing values and NO_2 hourly concentrations from the 200th to the 253rd hour (i.e., from 8:00 of the 9th of January 2000 to 13:00 of the 11th of January 2000)

Firstly, six consecutive NO_2 missing values from the 224th to the 229th hour of the year, i.e. from 8:00 to 13:00 of the 10th of January 2000, have been considered. Kriging hourly estimations for such missing values have been obtained using, alternatively

1. the variogram model (3.10), which describes the temporal correlation for NO_2 hourly concentrations,
2. the variogram model (3.11), which describes the temporal correlation for NO_2 hourly residuals.

In the former case, NO_2 hourly measurements have been directly estimated by the *GSLib* routine "KT3DP" properly modified in order to consider an ad hoc neighborhood, as above discussed. In the latter case, the original version of "KT3D" has been applied to estimate NO_2 residuals, then the diurnal component, previously estimated by the moving average technique, has been added to the estimated residuals, in order to obtain estimates of NO_2 hourly concentrations.

Figure 3.4 shows the time series of the estimated missing values (both estimated values obtained with the periodic variogram model and those obtained with nonperiodic variogram model), together with the time series of true NO_2 values corresponding to the period ranging from the 200th hour (i.e., 8:00 of the 9th of January 2000) to the 253rd hour (i.e., 13:00 of the 11th of January 2000).

Table 3.1 Kriging estimation of a sequence of six missing values

Hour	NO$_2$ obs. value	NO$_2$ est. value[a]	Est. std. dev.[a]	Est. error[a]	Est. resid.[b]	Est. std. dev.[b]	Trend[b]	Est. value[b]	Est. error[b]
224	46.2	44.026	10.264	−2.174	52.962	7.337	−5.835	47.127	0.927
225	53.1	56.915	18.543	3.815	58.335	9.271	−4.001	54.334	1.234
226	62.4	62.319	18.543	−0.081	63.578	10.048	−1.705	61.873	−0.527
227	71.9	74.344	18.543	2.444	68.234	10.049	1.743	69.977	−1.923
228	83.3	80.423	18.543	−2.877	73.787	9.272	7.544	81.331	−1.969
229	100.7	100.233	10.264	−0.467	79.712	7.337	16.189	95.901	−4.799

[a]Results obtained by using the periodic variogram model (3.10).
[b]Results obtained by using the nonperiodic variogram model (3.11).

The linear correlation coefficients between the NO$_2$ hourly true values and the estimates computed with the two different procedures confirm the validity of the estimation procedures. Moreover, the kriging standard deviation associated with each predicted missing value is lower if the nonperiodic variogram model is used, with respect to the case of kriging based on the periodic variogram model (Table 3.1). It is clear the flexibility of kriging to reconstruct the time series even when the periodic component is not factored out and the temporal correlation is described by a periodic variogram model.

Finally, three other sequences of missing values longer than the previous one have been considered. In fact, starting from the shortest sequence of six missing values above discussed, sequences of 12 (from the 224th to the 235th hour), 18 (from the 224th to the 241st hour), and 24 (from the 224th to the 247th hour) consecutive missing values have been assumed for the time series under study.

Table 3.2 reports the main results concerning the estimation of NO$_2$ missing values for the four intervals of time. It is clear that when kriging is performed for the original time series, the estimates of NO$_2$ values obtained with "KT3DP" using the periodic variogram model (3.10) keep on being characterized by high correlation coefficients even when the sequences of consecutive missing values are longer. On the other hand, when kriging is performed for the time series of NO$_2$ residuals, more consecutive missing values there are, more unsatisfactory results are produced. Note that, in this last case, there is a further element of bias due to the diurnal component which has been added to the estimated residual values in order to compute the estimates of NO$_2$ concentrations.

3.5.3 Prediction of NO$_2$ Values

In this section, the flexibility and the usefulness of geostatistical techniques, based on the variogram, have been exploited to make predictions for the variable under study. Hence, the above discussed variogram models (3.10) and (3.11) of NO$_2$ concentrations and NO$_2$ residuals, respectively, have been used in order to obtain

Table 3.2 Kriging estimations of 6, 12, 18, and 24 consecutive missing values

Interval of time	NO_2 avg. value	NO_2 avg. estimate[a]	Correlation coefficient[a]	NO_2 avg. estimate[b]	Correlation coefficient[b]
224–229	69.600	69.710	0.992	68.424	0.999
224–235	77.342	68.357	0.857	62.483	0.820
224–241	72.722	63.814	0.879	59.690	0.587
224–247	68.325	60.711	0.901	61.966	0.079

[a]Results obtained by using the periodic variogram model (3.10).
[b]Results obtained by using the nonperiodic variogram model (3.11).

kriging predictions for six intervals of time after the 31st of January 2000. In particular, the six different prediction periods have concerned intervals of:

- four hours (from the 745th to the 748th hour),
- eight hours (from the 745th to the 752th hour),
- twelve hours (from the 745th to the 756th hour),
- sixteen hours (from the 745th to the 760th hour),
- eighteen hours (from the 745th to the 762th hour),
- twenty-four hours (from the 745th to the 768th hour).

In other words, NO_2 hourly concentrations have been firstly predicted for a short period of 4 h and successively for longer sequences of time points, after the last available observation. These predictions have been obtained by using, alternatively

1. the available data, the variogram model (3.10) and the modified *GSLib* routine "KT3DP" which builds the searching neighborhood taking into account the periodicity exhibited by the data,
2. the deseasonalized NO_2 observations, the variogram model (3.11) and the original *GSLib* routine "KT3D" which produces NO_2 predicted residuals at which the diurnal component of the day before has been added to obtain predictions of NO_2 hourly concentrations.

Clearly, in case of extrapolation, the time points considered in the kriging system are just the observations available in the past.

In Fig. 3.5, the time series of NO_2 hourly concentrations measured from the 700th hour to the 748th hour are shown together with the predicted NO_2 values for the first interval, from the 745th to the 748th hour. Although the linear correlation coefficients between predictions and true values are satisfactory in both cases (0.999 and 0.988, respectively), the kriging procedure using the variogram model (3.11) related to NO_2 residuals produced overestimates of the pollution levels. This result could be due to the estimated diurnal component, which represents a further element of bias. Moreover, the kriging standard deviation associated with each predicted value is lower if the nonperiodic variogram model is used, with respect to the case of kriging based on the periodic variogram model (Table 3.3).

As mentioned above, the analysis has been developed to predict NO_2 values over intervals of time longer than four time points. Obviously, the longer is the predicted period, the greater is the inaccuracy of the predictions, especially when kriging has

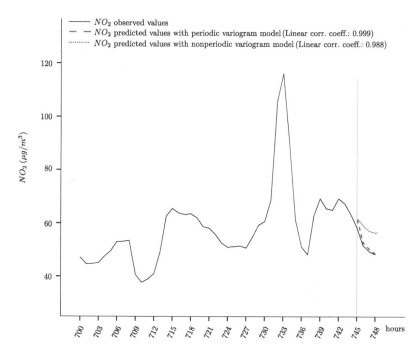

Fig. 3.5 Time plot of NO_2 predicted values and NO_2 hourly concentrations measured from the 700th to the 748th hour (i.e., from 4:00 of the 30th of January to 4:00 of the 1st of February 2000)

Table 3.3 Kriging predictions for 4 h

Hour	NO_2 obs. value	NO_2 pred. value[a]	Pred. std. dev.[a]	Pred. error[a]	Pred. resid.[b]	Pred. std. dev.[b]	Trend[b]	Pred.[b]	Pred. error[b]
745	58.3	62.058	10.339	3.758	62.650	7.665	−0.681	61.969	3.669
746	51.7	52.755	20.786	1.055	62.079	10.217	−2.853	59.226	7.526
747	49.5	50.448	20.786	0.948	61.622	11.862	−4.188	57.434	7.934
748	48.3	48.572	20.786	0.272	61.256	13.051	−4.628	56.628	8.328

[a]Results obtained by using the periodic variogram model (3.10).
[b]Results obtained by using the nonperiodic variogram model (3.11).

been performed using the deseasonalized NO_2 values and the nonperiodic variogram model (3.11). This is immediately confirmed by the results summarized in Table 3.4, concerning six different prediction periods, i.e. intervals of 4 (from the 745th to the 748th hour), 8 (from the 745th to the 752th hour), 12 (from the 745th to the 756th hour), 16 (from the 745th to the 760th hour), 18 (from the 745th to the 762th hour) and 24 (from the 745th to the 768th hour) units time. The linear correlation coefficients between true values and predicted ones decrease as the period to be predicted is longer, especially if the intervals of time are longer than 12 h (from 745th to 760th hour, from 745th to 762th hour, from 745th to 768th hour).

Table 3.4 Kriging predictions at 4, 8, 12, 16, 20, and 24 time points after the last available data

Interval of time	NO₂ avg. value	NO₂ avg. prediction[a]	Correlation coefficient[a]	NO₂ avg. prediction[b]	Correlation coefficient[b]
745–748	51.950	53.458	0.999	58.814	0.988
745–752	51.750	51.331	0.892	56.990	0.700
745–756	54.675	55.298	0.876	59.413	0.841
745–760	61.556	55.809	0.515	59.293	0.360
745–762	64.250	56.828	0.514	60.098	0.431
745–768	64.729	58.000	0.512	60.214	0.437

[a]Results obtained by using the periodic variogram model (3.10).
[b]Results obtained by using the nonperiodic variogram model (3.11).

3.5.4 Estimation of c.d.f.

The c.d.f. of NO_2 at unsampled time points has been estimated by indicator kriging [25].

In particular, six threshold values for NO_2 (36.30, 46.40, 50.80, 56.50, 67.28, and 79.12 $\mu g/m^3$) have been properly chosen and the observed values have been codified into indicator data (equal to 1 if the value is not greater than the threshold, 0 otherwise).

On the left-hand side of Figs. 3.6 and 3.7, the indicator time series, corresponding to the above-mentioned thresholds, has been shown through posting maps of hours of the day towards the days of January 2000.

Temporal indicator variogram has been computed for each threshold (left-hand side of Figs. 3.6 and 3.7), and the following models have been fitted:

$$\gamma_I(h_t;\ 36.30) = 0.02\,\mathrm{Exp}(|h_t|/6) + 0.16\,\mathrm{Exp}(|h_t|/70),$$

$$\gamma_I(h_t;\ 46.40) = 0.085\,\mathrm{Exp}(|h_t|/6) + 0.143\,\mathrm{Exp}(|h_t|/120) + 0.022\,\mathrm{Cos}(|h_t|/24),$$

$$\gamma_I(h_t;\ 50.80) = 0.11\,\mathrm{Exp}(|h_t|/6) + 0.12\,\mathrm{Exp}(|h_t|/110) + 0.02\,\mathrm{Cos}(|h_t|/24),$$

$$\gamma_I(h_t;\ 56.50) = 0.042\,\mathrm{Exp}(|h_t|/6) + 0.067\,\mathrm{Exp}(|h_t|/20) + 0.094\,\mathrm{Exp}(|h_t|/110)$$
$$+ 0.012\,\mathrm{Cos}(|h_t|/12) + 0.024\,\mathrm{Cos}(|h_t|/24),$$

$$\gamma_I(h_t;\ 67.28) = 0.097\,\mathrm{Exp}(|h_t|/12) + 0.034\,\mathrm{Exp}(|h_t|/60) + 0.009\,\mathrm{Cos}(|h_t|/12)$$
$$+ 0.017\,\mathrm{Cos}(|h_t|/24),$$

$$\gamma_I(h_t;\ 79.12) = 0.002\,\mathrm{Exp}(|h_t|/3) + 0.074\,\mathrm{Exp}(|h_t|/12) + 0.006\,\mathrm{Cos}(|h_t|/12)$$
$$+ 0.01\,\mathrm{Cos}(|h_t|/24).$$

Then the "KT3DP" routine has been used to estimate the c.d.f. corresponding to three different unsampled time points, i.e. at hours 1:00, 10:00, and 19:00 of the 1st of February 2000. For each hour of interest, the c.d.f. has been estimated by solving as many kriging systems as the number of threshold values considered. For each threshold, the corresponding indicator variogram model has been used for the

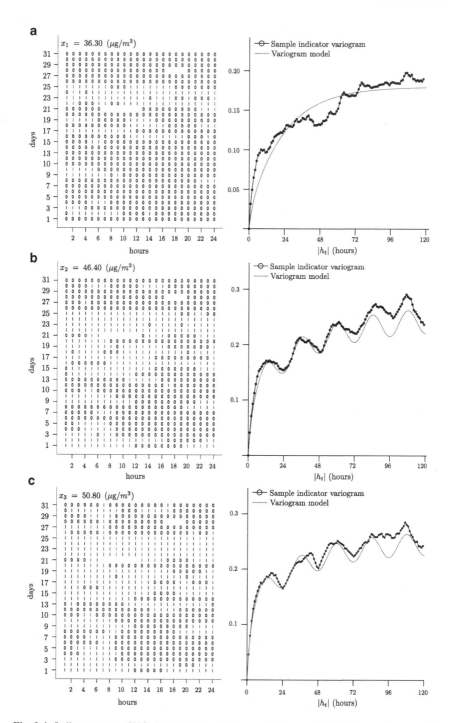

Fig. 3.6 Indicator maps of NO_2 hourly concentrations and their sample indicator variograms with models, for three threshold values. (**a**) Indicator map and variogram for the threshold $x_1 = 36.30$. (**b**) Indicator map and variogram for the threshold $x_2 = 46.40$. (**c**) Indicator map and variogram for the threshold $x_3 = 50.80 \ \mu g/m^3$

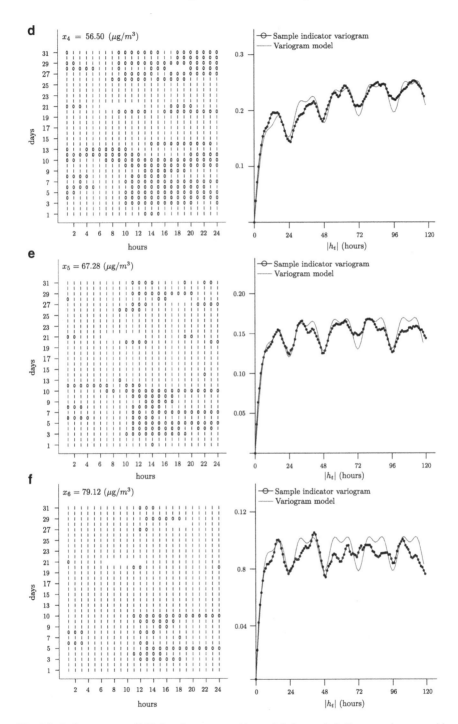

Fig. 3.7 Indicator maps of NO$_2$ hourly concentrations and their sample indicator variograms with models, for three threshold values. (**a**) Indicator map and variogram for the threshold $x_4 = 56.50$. (**b**) Indicator map and variogram for the threshold $x_5 = 67.28$. (**c**) Indicator map and variogram for the threshold $x_6 = 79.12$ $\mu g/m^3$

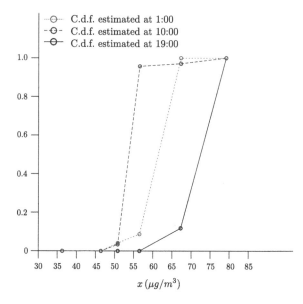

Fig. 3.8 C.d.f.s estimated for NO_2 hourly concentrations at 1:00, 10:00 and 19:00 of the 1st of February 2000

Table 3.5 Estimated values for c.d.f. at hours 1:00, 10:00, and 19:00 of the 1st of February 2000, for fixed thresholds

Threshold	Hour 1:00	Hour 10:00	Hour 19.00
36.30	0.000	0.000	0.000
46.40	0.000	0.000	0.000
50.80	0.041	0.034	0.000
56.50	0.087	0.957	0.000
67.28	1.000	0.971	0.118
79.12	1.000	1.000	1.000

kriging procedure. Figure 3.8 shows the c.d.f.s estimated at hours 1:00, 10:00, and 19:00 of the 1st of February 2000.

It is clear that the probability of not exceeding a fixed threshold reduces gradually over the day. For example, the estimated probability that NO_2 concentrations, on the 1st of February 2000, do not exceed 56.50 $\mu g/m^3$ is higher at 10:00 than in the evening (19:00) and in the night (1:00). Moreover, note that it is sure or almost sure that NO_2 concentrations do not exceed the cutoff 67.28 $\mu g/m^3$ at 1:00 and 10:00; on the other hand, it is not likely that NO_2 levels do not exceed 67.28 $\mu g/m^3$ at 19:00 (Table 3.5).

This result might be useful for the local government in fixing directives to protect and control the air quality, in order to avoid the risk of high levels of nitrogen dioxide pollution, which might be dangerous for the ecosystem and the human health.

3.6 Conclusions

In this paper, the potentiality of geostatistical techniques and its basic tool, such as the variogram, in time series analysis, have been pointed out.

Although in literature the use of the variogram in time series analysis has been highlighted in different ways, as widely discussed in this paper, time series analysts do not typically use Geostatistics for exploratory analysis or prediction, even when an explicit model for the process is not so relevant. Moreover, most of the software for time series analysis does not provide variogram-based techniques in order to face and solve inferential problems.

The importance and convenience of Geostatistics to perform a complete analysis of a time series have been underlined through the case study, where the exploratory and prediction stages, even in presence of a periodic behavior, have been presented and useful computational aspects have been highlighted. A modified version of the *GSLib* routine for kriging has been also implemented for missing values treatment and prediction of time series with a periodic component.

As underlined in [53], time series analysis will keep on representing one of the main appealing fields of the scientific research, in the presence of lively interaction with other fields and more and more efforts in several directions. Among these, a few steps further in using geostatistical tools in time series analysis are expected, especially regarding software implementation.

Acknowledgements The research activities involved in this paper were partially supported by the Project "5 per mille per la ricerca" entitled "Modelli di Interpolazione Stocastica per il Monitoraggio Ambientale: Sviluppi Teorici e Applicativi" given by University of Salento for the period 2011–2012 (scientific coordinator prof. S. De Iaco).

References

1. Alabert, F.: The practice of fast conditional simulations through the LU decomposition of the covariance matrix. Math. Geol. **19**(5), 369–386 (1987)
2. Banea, O., Solow, A.R., Stone, L.: On fitting a model to a population time series with missing values. Isr. J. Ecol. Evol. **52**(1), 1–10 (2006)
3. Bisgaard, S., Khachatryan, D.: Asymptotic confidence intervals for variograms of stationary time series. Qual. Reliab. Eng. Int. (2007). doi:10.1002/qre.1052
4. Bloomfield, P.: Fourier Analysis of Time Series: An Introduction, 2nd Edition J. Wiley & Sons, Inc., USA (2000)
5. Box, G.E.P., Jenkins, G.M.: Time Series Analysis: Forecasting and Control. Holden Day, San Francisco (1976)
6. Brockwell, P.J., Davis, R.A.: Time Series: Theory and Methods. Springer, New York (1987)
7. Chilés, J.P., Delfiner, P.: Geostatistics: Modeling Spatial Uncertainty. Wiley, New York (1999)
8. Christakos, G.: Modern Spatiotemporal Geostatistics. Oxford University Press, New York (2000)

9. Cressie, N.: A graphical procedure for determining nonstationarity in time series. J. Am. Stat. Assoc. **83**(404), 1108–1116 (1988)
10. Cressie, N.: Statistics for Spatial Data, Wiley Series in Probability and Mathematical Statistics. Wiley, New York (1993)
11. Cressie, N., Grondona, M.O.: A comparison of variogram estimation with covariogram estimation. In: Mardia, K.V. (ed.) The Art of Statistical Science. Wiley, Chichester (1992)
12. Cressie, N., Wikle, C.K.: Statistics for Spatio-Temporal Data. J. Wiley & Sons, Inc., Hoboken, New Jersey (USA) (2011)
13. Deutsch, C.V., Journel, A.G.: *GSLib*: Geostatistical Software Library and User's Guide. Oxford University Press, New York (1998)
14. De Cesare, L., Myers, D.E., Posa, D.: FORTRAN 77 programs for space-time modeling. Comput. Geosci. **28**(2), 205–212 (2002)
15. De Iaco, S., Myers, D.E., Posa, D.: On strict positive definiteness of product and product-sum covariance models. J. Stat. Plan. Inf. **141**(3), 1132–1140 (2011)
16. De Iaco, S., Myers, D.E., Posa, D.: Strict positive definiteness of a product of covariance functions. Commun. Stat. Theory Methods **40**(24), 4400–4408 (2011)
17. De Iaco, S., Posa, D.: Predicting spatio-temporal random fields: some computational aspects. Comput. Geosci. **41**, 12–24 (2012)
18. De Iaco, S., Palma, M.: Convergence of realization-based statistics to model-based statistics for the LU unconditional simulation algorithm. Some numerical tests. Stoch. Environ. Res. Risk Assess. Springer **16**(5), 333–341 (2002)
19. Gandin, L.S.: Objective Analysis of Meteorological Fields. Gidrometeorologicheskoe Izdatelstvo, Leningrad (1963)
20. Gevers, M.: On the use of variograms for the prediction of time series. Syst. Control Lett. **6**(1), 15–21 (1985)
21. Gomez-Hernandez, J.J., Cassiraga, E.F.: Theory and practice of sequential simulation. In: Armstrong, M., Dowd, P.A. (eds) Geostat. Simul., pp. 111–124. Kluwer Academic, Norwell, Massachusetts (1994)
22. Harris, B.: Introduction to the Theory of Spectral Analysis of Time Series in Spectral Analysis of Time Series. Wiley, New York (1967)
23. Haslett, J.: On the sample variogram and sample autocovariance for non-stationary time series. Statistician **46**(4), 475–485 (1997)
24. Hull, J., White, A.: The pricing of assets on options with stochastic volatilities. J. Finance **42**(2), 281–300 (1987)
25. Janis, M.J., Robeson, S.M.: Determining the spatial representativeness of air-temperature records using variogram-nugget time series. Phys. Geogr. **25**(6), 513–530 (2004)
26. Jenkins, G.M., Watts, D.G.: Spectral Analysis and Its Applications. Holden-Day, New York (1968)
27. Jones, R.H., Ackerson, L.M.: Serial correlation in unequally spaced longitudinal data. Biometrika **77**(4), 721–732 (1990)
28. Journel, A.G.: Nonparametric estimation of spatial distributions. Mat. Geol. **15**(3), 445–468 (1983)
29. Journel, A.G., Huijbregts, C.J.: Mining Geostatistics. Academic, London (1981)
30. Journel, A.G., Rossi, E.M.: When do we need a trend model in kriging? Mat. Geol. **21**(7), 715–739 (1989)
31. Khachatryan, D., Bisgaard, S.: Some results on the variogram in time series analysis. Qual. Reliab. Eng. Int. (2009). doi:10.1002/qre.1013
32. Kolmogorov, A.N.: The local structure of turbulence in an incomprehensible fluid at very large Reynolds numbers. Dokl. Acad. Nauk. SSSR **30**(4), 229–303 (1941)
33. Kyriakidis, P.C., Journel, A.G.: Geostatistical space-time models: a review. Mat. Geol. **31**(6), 651–684 (1999)
34. Krige, D.G.: A statistical approach to some basic mine valuation problems on the Witwatersrand. J. Chem. Metall. Min. Soc. S. Afr. **52**(6), 119–139 (1951)

35. Little, R.J.A., Rubin, D.B.: Statistical Analysis with Missing Data. Wiley, New York (2002)
36. Luceño, A.: Estimation of missing values in possibly partially nonstationary vector time series. Biometrika **84**(2), 495–499 (1997)
37. Ma, C.: The use of the variogram in construction of stationary time series models. J. Appl. Probab. **41**(4), 1093–1103 (2004)
38. Ma, C.: Stochastic processes with a particular type of variograms. Res. Lett. Signal Process (2007). doi:10.1155/2007/61579
39. Ma, C.: Recent developments on the construction of spatio-temporal covariance models. Stoch. Environ. Res. Risk Assess. **22**(S1), 39–47 (2008)
40. Matheron, G.: Principles of Geostatistics. Econ Geol **58**(8), 1246–1266 (1963)
41. Matheron, G.: Les variables régionalisées et leur estimation. Masson, Paris (1965)
42. Matheron, G.: The intrinsic random functions and their applications. Adv. Appl. Probab. **5**(3), 439–468 (1973)
43. Micchelli, C.: Interpolation of scattered data: distance matrices and conditionally positive definite functions. Constr. Approx. **2**, 11–22 (1986)
44. Myers, D.E.: Interpolation with positive definite functions. Sci. Terre **28**, 251–265 (1988)
45. Myers, D.E.: Kriging, cokriging, radial basis functions and the role of positive definiteness. Comput. Math. Appl. **24**(12), 139–148 (1992)
46. Myers, D.E., De Iaco, S., Posa, D., De Cesare, L.: Space-time radial basis functions. Comput. Math. Appl. **43**(3–5), 539–549 (2002)
47. Posa, D.: Limiting stochastic operations for stationary spatial processes. Math. Geol. **23**(5), 695–701 (1991)
48. Posa, D.: The indicator formalism in spatial data analysis. J. Appl. Stat. **19**(1), 83–101 (1992)
49. Posa, D., Rossi, M.: Applying stationary and non-stationary kriging. Metron **XLVII**(1–4), 295–312 (1989)
50. Posa, D., De Iaco, S.: Geostatistica: teoria e applicazioni. Giappichelli editore, Torino (2009)
51. Priestley, M.B.: Spectral Analysis and Time Series. Academic, London (1981)
52. Schoellhamer, D.H.: Singular spectrum analysis for time series with missing data. Geophys Res Lett **28**(16), 3187–3190 (2001)
53. Schoenberg, I.J.: Metric spaces and positive definite functions. Trans. Am. Mat. Soc. **44**(3), 522–536 (1938)
54. Solow, A.R.: The analysis of second-order stationary processes: times series analysis, spectral analysis, harmonic analysis, and geostatistics. In: Verly, G., et al. (eds.) Geostatistics for Natural Resources Characterization, Part I, pp. 573–585. D. Reidel Publishing Co, Dordrecht (1984)
55. Stein, M.L.: The screening effect in kriging. Ann. Stat. **30**(1), 298–323 (2002)
56. Tong, H.: A personal journey through time series in Biometrika. Biometrika **88**(1), 195–218 (2001)
57. Uysal, M.: Reconstruction of time series data with missing values. J. Appl. Sci. **7**(6), 922–925 (2007)
58. Weerasinghe, S.: A missing values imputation method for time series data: an efficient method to investigate the health effects of sulphur dioxide levels. Environmetrics (2009). doi:10.1002/env.990
59. Yaglom, A.M.: Correlation Theory of Stationary and Related Random Functions. Springer, New York (1987)
60. Yaglom, A.M.: An Introduction to the Theory of Stationary Random Functions. Dover, New York (2004)

Chapter 4
GIS and Geostatistics for Supporting Environmental Analyses in Space-Time

Sabrina Maggio, Claudia Cappello, and Daniela Pellegrino

Abstract The environmental risk assessment involves the analysis of complex phenomena. Different kinds of information, such as environmental, socio-economic, political and institutional data, are usually collected. In this chapter, spatio-temporal geostatistical analysis is combined with the use of a Geographic Information System (*GIS*): the integration between geostatistical tools and *GIS* enables the identification and the visualization of alternative scenarios regarding a phenomenon under study and supports the environmental risk management.

A case study on environmental data measured at different monitoring stations in the southern part of Apulia Region (South of Italy), called Grande Salento, is discussed. Sample data concerning daily averages of PM_{10}, Wind Speed and Atmospheric Temperature, are used for stochastic prediction, through space–time indicator kriging.

Kriging results are implemented in a *GIS* and a 3D representation of the spatio-temporal probability maps is proposed.

Keywords Conditional probability map • Geostatistics • *GIS* • PM_{10} pollution • Space–time indicator kriging

4.1 Introduction

Environmental risk management involves the integrated use of several tools and techniques, including Geographic Information System (*GIS*), Geostatistics and data management. In particular, a data management process requires the integration of several data, which are usually classified into three categories: (1) environmental

S. Maggio (✉) • C. Cappello • D. Pellegrino
Department of Management and Economics, University of Salento, Complesso Ecotekne, Via per Monteroni, 73100 Lecce, Italy
e-mail: sabrina.maggio@unisalento.it; claudia.cappello@unisalento.it; d.pellegrino83@libero.it

S. Montrone and P. Perchinunno (eds.), *Statistical Methods for Spatial Planning and Monitoring*, Contributions to Statistics, DOI 10.1007/978-88-470-2751-0_4, © Springer-Verlag Italia 2013

data (land use, land cover, vegetation, geology, meteorology and measures of pollutant concentrations); (2) socio-economic data (population and housing census data, community vulnerability data and data on utilities and access); (3) political and institutional data [5].

GIS consists of different tools for storage, retrieval, analysis and display of spatial data and, in some cases, of spatio-temporal data. Thus, it could be a valid support for modelling and prediction purposes. However, a thorough geostatistical analysis for spatial and spatio-temporal data can be provided by using appropriate geostatistical techniques [24] which are not implemented in the most used GIS software.

Hence, the interaction of space–time modelling and prediction geostatistical techniques [10, 19, 23, 25] with urban environment representation (traffic network, location of industrial facilities, emission sources and topographic conditions), easily managed in a GIS, is necessary.

The problem of integration between GIS and Geostatistics has been addressed since early 1990s when Goodchild [18] illustrated the potential advantages of an integrated use of GIS and spatial analysis. Accordingly, over the years, many researchers have approached this problem in different ways [1, 3, 4].

The aim of the chapter is to combine the use of space–time geostatistical techniques and the GIS potential for environmental studies in order to analyse a dangerous pollutant for the human health.

Hence, after a brief introduction on the integration between GIS and Geostatistics (Sect. 4.2), a review of the spatio-temporal geostatistical techniques for modelling and non-parametric prediction purposes (Sect. 4.3) is presented. Finally, a case study based on an environmental data set is proposed (Sect. 4.4). The analysed data involve two atmospheric variables (wind speed and temperature) and particulate matter concentrations, measured in November 2009 at some monitoring stations located in the Grande Salento (the districts of Lecce, Brindisi and Taranto in the Apulia Region, Italy). In particular, the variables under study regards: (1) PM_{10} (particulate matter with diameter smaller than 10 mm) daily averages concentrations, because of their negative effects on human health, (2) wind speed (WS) and atmospheric temperature (AT) daily averages, because of their significant role on the particle pollution.

Exploratory spatial data analysis for a deep understanding of the analysed phenomenon has been performed using the Geostatistical Analyst Tool of ArcGIS. Space–time indicator variogram modelling has been provided and spatio-temporal indicator kriging has been computed by using some modified GSLib routines. Conditional probabilities that PM_{10} concentrations are not greater than fixed thresholds when the atmospheric variables (WS and AT) are lower than the corresponding monthly means have been determined.

Three-dimensional representations for the space–time evolution of above-mentioned conditional probability associated with PM_{10} have been produced by using ArcScene (an extension of ArcGIS).

4.2   GIS and Geostatistics

In recent years, many environmental studies have been conducted by using *GIS* and Geostatistics techniques jointly [16, 20, 22].

In this context, *GIS* enables researchers to link results from geostatistical analysis with some geographic information such as land use, traffic networks and industrialized areas. On the other hand, Geostatistics provides advanced techniques to predict a variable of interest at unsampled points and to support decisions concerning monitoring, sampling, planning and requalification of the territory [6].

Hence, different methods of spatial analysis have been implemented in several kinds of *GIS* softwares, such as variogram estimation and modelling, ordinary kriging and indicator kriging in *Geostatistical Analyst Tool* of ArcGIS.

Moreover, the open source software *GRASS* (*Geographic Resources Analysis Support System*) interfaces with R, by means of the *spgrass6* module, in order to enable a geostatistical data analysis.

Up to now, the most remarkable developments concern the integration between *GIS* and geostatistical tools for the analysis of spatio-temporal data. For example, *ArcScene*, which is an extension of *ArcGIS*, allows to obtain 3D representations in a *GIS* project.

4.3   Geostatistical Framework

Environmental risk analysis requires the observation of several variables characterized by space–time evolution. Hence, either classical multivariate methods and space–time geostatistical techniques might be applied to analyse, interpret and control the complex behavior of the observed variables [8].

In this context, the observations of each variable are modelled as a realization of a second order stationary spatio-temporal random function (*STRF*)

$$\{Z(\mathbf{u}), \mathbf{u} = (\mathbf{s}, t) \in D \times T \subseteq \mathbb{R}^2 \times \mathbb{R}\},$$

with the following first and second moments

- $E[Z(\mathbf{s}, t)] = m,$
- $\text{Cov}[Z(\mathbf{s} + \mathbf{h}_s, t + h_t), Z(\mathbf{s}, t)] = C(\mathbf{h}_s, h_t),$
- $\text{Var}[Z(\mathbf{s} + \mathbf{h}_s, t + h_t) - Z(\mathbf{s}, t)] = 2\gamma(\mathbf{h}_s, h_t).$

Given a second order stationary space–time random field Z, for a fixed threshold $z \in \mathbb{R}$, a spatio-temporal indicator random field (*STIRF*)

$$\{I(\mathbf{u}, z), \mathbf{u} = (\mathbf{s}, t) \in D \times T\} \tag{4.1}$$

is defined as follows:

$$I(\mathbf{u}, z) = \begin{cases} 1 & \text{if } Z \text{ is not greater (or not smaller) than the threshold } z, \\ 0 & \text{otherwise.} \end{cases}$$

Spatio-temporal correlation of a stationary *STIRF* is described by the indicator variogram:

$$2\gamma_I(\mathbf{h}; z) = \mathrm{Var}[I(\mathbf{u}+\mathbf{h}; z) - I(\mathbf{u}; z)],$$

which depends on the threshold z and the lag vector $\mathbf{h} = (\mathbf{h}_s, h_t)$, with $(\mathbf{s}, \mathbf{s} + \mathbf{h}_s) \in D^2$ and $(t, t + h_t) \in T$.

The fitted model for $\gamma_I(\cdot; \cdot)$ must satisfy an admissibility condition in order to be valid.

Among different spatio-temporal models proposed in literature [7, 11, 15, 17, 19, 21], the generalized product–sum model [10] can be easily fitted to the empirical spatio-temporal variogram of the *STIRF* (4.1). This model is properly defined as follows:

$$\gamma_I(\mathbf{h}_s, h_t; z) = \gamma_I(\mathbf{h}_s, 0; z) + \gamma_I(\mathbf{0}, h_t; z) - k\gamma_I(\mathbf{h}_s, 0; z)\gamma_I(\mathbf{0}, h_t; z), \qquad (4.2)$$

where

- $\gamma_I(\mathbf{h}_s, 0; z)$ is a valid spatial bounded marginal variogram.
- $\gamma_I(\mathbf{0}, h_t; z)$ is a valid temporal bounded marginal variogram.
- $k \in]0, 1/max\{sill_{\gamma_I}(\mathbf{h}_s, 0; z), sill_{\gamma_I}(\mathbf{0}, h_t; z)\}]$ is the parameter of spatio-temporal interaction.

A comparative analysis and further theoretical results can be found in [10]; moreover, recently it has been shown that strict conditional negative definiteness of both marginals is a necessary and a sufficient condition for the product–sum (4.2) to be strictly conditionally negative definite [12, 13].

Indicator kriging allows the estimation of the probability of exceeding specific threshold values z, at a given location. At an unsampled point of the domain of interest, the probability that Z is not greater (or not smaller) than the threshold z can be estimated using a linear combination of neighbouring indicator variables. The indicator kriging estimator is defined as follows:

$$\widehat{I}(\mathbf{u}; z) = \sum_{\alpha=1}^{n} \lambda_\alpha(\mathbf{u}_\alpha; z) I(\mathbf{u}_\alpha; z)$$

where $\mathbf{u}$ is an unsampled point, $I(\mathbf{u}_\alpha; z), \alpha = 1, 2, \ldots, n$ represent the indicator random variables at the sampled points $\mathbf{u}_\alpha \in D \times T$ and $\lambda_\alpha(\mathbf{u}_\alpha; z)$ are the kriging weights which are determined by solving the following indicator kriging system:

$$\begin{cases} \sum_{\beta=1}^{n} \lambda_\beta(\mathbf{u}; z)\gamma_I(\mathbf{u}_\alpha - \mathbf{u}_\beta; z) - \mu(\mathbf{u}; z) = \gamma_I(\mathbf{u}_\alpha - \mathbf{u}; z), \quad \alpha = 1, 2, ..., n \\ \sum_{\alpha=1}^{n} \lambda_\alpha(\mathbf{u}; z) = 1 \end{cases}$$

where μ is the Lagrange multiplier.

4.4 Case Study

In the present case study, PM_{10} concentrations, WS and AT have been analysed and stochastic prediction for PM_{10}, through space–time indicator kriging, has been proposed.

Particulate matter (PM) is a complex mixture of organic and inorganic substances, such as sulphates, nitrates, ammonia, sodium chloride, carbon, mineral dust, water and metal. There are several kinds of particles, differentiated by size, composition and origin. In particular, PM_{10} is composed by particles with a diameter smaller than 10 μm.

Note that the size of the particles determines the time they spend in the atmosphere: sedimentation and precipitations remove PM_{10} from the atmosphere within a few hours after the emission, consequently it cannot be transported. Moreover, high values of WS and low values of AT affect the pollutant dispersion.

The study of PM_{10} evolution is very important for its effects on human health. Many medical researches have shown that exposure to PM_{10} increases the risk of mortality both in long and short term; for example, it has been demonstrated the existence of correlation between PM_{10} concentrations and the presence of chronic respiratory disease [2].

4.4.1 The Data Set

The environmental data set consists of PM_{10} ($\mu g/m^3$), WS (m/s) and AT ($°C$) daily averages, measured in November 2009 at 28 monitoring stations located in the Grande Salento (the districts of Lecce, Brindisi and Taranto, in the Apulia Region, Italy), as shown in Fig. 4.1.

Data concerning PM_{10} concentrations are provided by "ARPA Puglia", while atmospheric data by "ASSOCODIPUGLIA".

PM_{10} survey stations are classified in the following three categories:

1. traffic stations, located in areas with heavy traffic,
2. industrial stations, located close to industrialized areas,
3. ground stations, located in peripheral areas.

Exploratory spatial data analysis has been performed by using *Geostatistical Analyst Tool* of *ArcGIS*. The statistical properties of PM_{10}, WS and AT have been assessed. Some results are shown in Table 4.1.

According to National Laws concerning the human health protection, PM_{10} daily average concentrations cannot be greater than 50 $\mu g/m^3$ for more than 35 times per year. During the month under study, the PM_{10} daily values exceeded the threshold 80 times, especially on the 13th, 14th, 23th and 24th of November (Fig. 4.2). Daily averages of the variables under study have been used for stochastic prediction of the PM_{10} concentrations during the period 1–6 December 2009, through space–time indicator kriging.

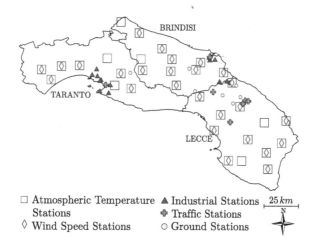

Fig. 4.1 Location map of PM$_{10}$, WS and AT survey stations, located in the Grande Salento

Table 4.1 Descriptive statistics of PM$_{10}$ ($\mu g/m^3$), WS (m/s) and AT ($^{\circ}C$) values, measured in November 2009 in the Grande Salento

	Min.	Max.	Mean	Standard deviation	75th percentile	80th percentile
PM$_{10}$	4.477	110.10	30.81	15.973	37.804	40.57
WS	0.414	8.893	2.11	1.340	2.61	2.786
AT	8.158	17.995	12.53	2.062	13.818	14.105

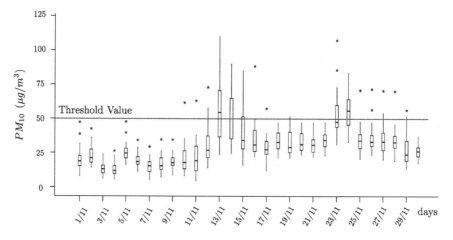

Fig. 4.2 Box plots of PM$_{10}$ daily average concentrations in November 2009, classified by days

In the present case study, the following aspects have been considered.

1. Definition of the space–time indicator variables according to appropriate thresholds, computed from the observed data.
2. Estimating and modelling the space–time indicator variogram.
3. Using space–time indicator kriging, over the area of interest and during the period 1–6 December 2009, in order to obtain

    (a) the joint probability that PM_{10} concentrations exceed fixed thresholds and the atmospheric values (WS and AT daily averages) are not greater than the corresponding monthly means (adverse conditions for PM_{10} dispersion),
    (b) the joint probability that WS and AT daily averages are not greater than the corresponding monthly means.

4. Three-dimensional representations of the probability that PM_{10} concentrations exceed the fixed thresholds, conditioned to adverse atmospheric conditions for PM_{10} dispersion (i.e. WS and AT daily averages lower than the corresponding monthly mean values).

Note that *ArcScene*, an extension of *ArcGIS*, has been used in the paper to display the results obtained from the spatio-temporal analysis discussed in the following sections.

4.4.2   Structural Analysis

As previously pointed out, spatio-temporal observations for the variables under study have been considered as a realization of a stationary *STRF Z*.

The formalism of a *STIRF* (4.1) has been applied to the space–time data of PM_{10}, AT and WS. Thus, three spatio-temporal indicator variables have been defined on the basis of the following thresholds, for:

- PM_{10}, the 75th and 80th percentiles of samples data (37.804 and 40.57 $\mu g/m^3$, respectively), which can be considered critical values with respect to the law limit.
- AT and WS, the corresponding monthly mean values, i.e. 12.53 $°C$ and 2.11 m/s, respectively.

Spatio-temporal indicator kriging using the generalized product–sum variogram model (4.2) has been applied in order to predict, over the area of interest and for the period 1–6 December 2009, the probability that PM_{10} concentrations exceed the above-mentioned fixed limits, in the presence of adverse atmospheric conditions to the pollutant dispersion. Hence, three indicator random fields have been defined as follows.

1. Indicator random field I_1 is equal to 1 if PM_{10} daily average concentrations are not less than the threshold $z_1 = 37.804 \ \mu g/m^3$ (75th percentile) and in the presence of adverse atmospheric conditions to the pollutant dispersion, I_1 is equal to 0, otherwise. It is defined as follows:

$$I_1(\mathbf{u};\ 37.804;\ 12.53;\ 2.11) = \begin{cases} 1 & \text{if } PM_{10} \geq 37.804,\ AT \leq 12.53, \\ & WS \leq 2.11, \\ 0 & \text{otherwise,} \end{cases}$$

with $\mathbf{u} \in D \times T$.

2. Indicator random field I_2 is equal to 1 if PM_{10} daily average concentrations are not less than the threshold $z_2 = 40.57\ \mu g/m^3$ (80th percentile) and in the presence of adverse atmospheric conditions to the pollutant dispersion, I_2 is equal to 0, otherwise. It is defined as follows:

$$I_2(\mathbf{u};\ 40.57;\ 12.53;\ 2.11) = \begin{cases} 1 & \text{if } PM_{10} \geq 40.57,\ AT \leq 12.53, \\ & WS \leq 2.11, \\ 0 & \text{otherwise,} \end{cases}$$

with $\mathbf{u} \in D \times T$.

3. Indicator random field I_3 is equal to 1 if the atmospheric conditions are adverse to the pollutant dispersion, I_3 is equal to 0, otherwise. It is defined as follows:

$$I_3(\mathbf{u};\ 12.53;\ 2.11) = \begin{cases} 1 & \text{if } AT \leq 12.53,\ WS \leq 2.11, \\ 0 & \text{otherwise,} \end{cases}$$

with $\mathbf{u} \in D \times T$.

4.4.2.1   Estimating and Modelling

In order to model the spatio-temporal variogram surfaces of the indicator variables under study, using the generalized product–sum variogram model (4.2), the following steps have been faced:

- estimating sample space–time variograms and marginal variograms in space and time for the indicator variables I_1, I_2 and I_3,
- fitting marginal indicator variograms and identification of the sill values,
- determining the coefficient k of the generalized product–sum model.

Sample space–time variogram surfaces, sample marginal variograms for space and time and corresponding fitted models for the random fields I_1, I_2 and I_3 are shown in Fig. 4.3.

Graphical inspection of the spatio-temporal surfaces allows identifying the global sills needed to compute the parameters k.

In particular, the models fitted to the spatial and temporal marginal variograms, k parameters and global sills, are the following:

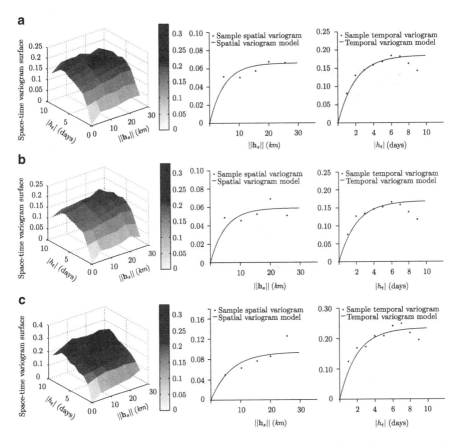

Fig. 4.3 Sample space–time variogram surfaces, sample marginal spatial and temporal variograms and relative fitted models for the indicator variables. (**a**) Indicator random field I_1. (**b**) Indicator random field I_2. (**c**) Indicator random field I_3

1. for *STIRF* I_1

   (a) $\gamma_I(\mathbf{h}_s, 0; z_1) = 0.066[1 - \exp(-3\|\mathbf{h}_s\|/15)]$;
   (b) $\gamma_I(\mathbf{0}, h_t; z_1) = 0.185[1 - \exp(-3h_t/6)]$;
   (c) $k = 3.767$;
   (d) Global sill equal to 0.205;

2. for *STIRF* I_2

   (a) $\gamma_I(\mathbf{h}_s, 0; z_2) = 0.059[1 - \exp(-3\|\mathbf{h}_s\|/15)]$;
   (b) $\gamma_I(\mathbf{0}, h_t; z_2) = 0.169[1 - \exp(-3h_t/6)]$;
   (c) $k = 4.112$;
   (d) Global sill equal to 0.187;

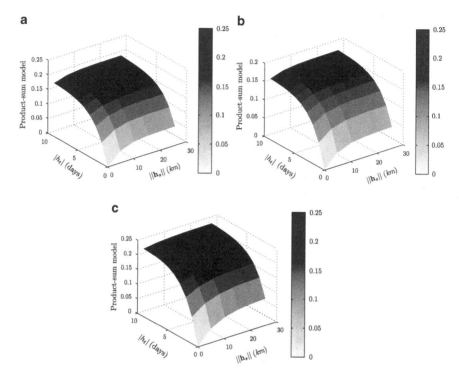

Fig. 4.4 Space–time variogram surfaces for the indicator variables. (**a**) Indicator random field I_1. (**b**) Indicator random field I_2. (**c**) Indicator random field I_3

3. for *STIRF* I_3

    (a) $\gamma_I(\mathbf{h}_s,\, 0;\, z_3) = 0.094[1 - \exp(-3\|\mathbf{h}_s\|/20)]$;

    (b) $\gamma_I(\mathbf{0}, h_t;\, z_3) = 0.235[1 - \exp(-3h_t/6)]$;

    (c) $k = 3.712$;

    (d) Global sill equal to 0.247.

Figure 4.4 displays the space–time variogram models for the indicator variables I_1, I_2 and I_3.

4.4.3 Space–Time Predictions Based on Spatio-Temporal Indicator Kriging

Finally, probability maps have been predicted over the area of interest, for the period 1–6 December 2009, by using some modified *GsLib* routines [12]. In particular, the indicator kriging has been used to estimate:

- the joint probability that PM_{10} concentrations exceed fixed thresholds and the daily averages of atmospheric variables are not greater than the corresponding monthly means,
- the joint probability that the daily averages of WS and AT are not greater than the corresponding monthly means.

Then, the conditional probabilities that PM_{10} values do not exceed the fixed thresholds (75th and 80th percentiles), under the established unfavourable weather conditions to pollutant dispersion, over the area of interest and during the period 1–6 December 2009, have been computed by using the probability values previously estimated.

The results, obtained by using *GSLib* modified routines, have been implemented in *ArcMap* by generating and storing, for each fixed threshold, a shapefile per day.

Hence, maps concerning the conditional probability that the pollutant concentrations exceed the established thresholds, in the presence of adverse atmospheric conditions to the pollutant dispersion, have been created by the implementation of above-mentioned shapefiles in a *GIS* project.

The maps of conditional probability for the first threshold $z_1 = 37.804 \, \mu g/m^3$ and the second threshold $z_2 = 40.57 \, \mu g/m^3$ are shown in Figs. 4.5 and 4.6, respectively. They are produced by using *ArcMap*. It is evident that the probability that PM_{10} daily concentrations exceed the fixed thresholds, under the established atmospheric conditions, is high in the central part of the area of interest.

In particular, it is high, during the estimated period, along the boundary among the districts of Lecce, Brindisi and Taranto, in the northern part (the district of Brindisi) and in the southern part (the district of Lecce) of the predicted area. This behaviour is due to the presence of pollution sources such as industrial establishments and heavy traffic, as well as to the occurrence of unfavourable atmospheric conditions to the pollutant dispersion.

Figure 4.7 shows a 3D representation (produced by *ArcScene* of *ArcGIS*) of the space–time evolution of the conditional probability associated with PM_{10}, with reference to the fixed thresholds.

4.5 Conclusions

In this chapter the integration of advanced geostatistical techniques into a *GIS* has been proposed. This integration has been realized by implementing the results of the spatio-temporal analysis, based on spatio-temporal indicator kriging, in a *GIS* project developed by using *ArcGIS* of *ESRI*.

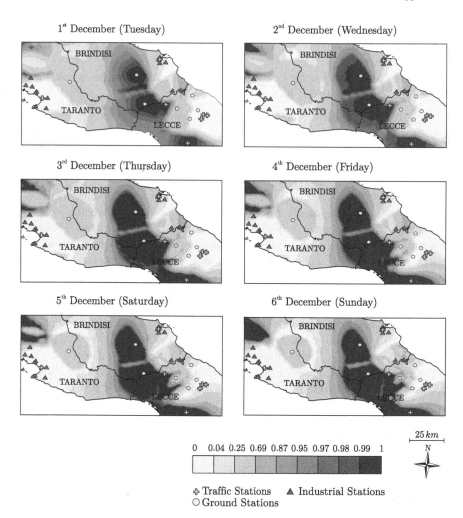

Fig. 4.5 Conditional probability maps of PM_{10} concentrations, for $z_1 = 37.804\,\mu g/m^3$ (75th percentile), in the Grande Salento, during the period 1–6 December 2009, obtained by using *ArcMap*

Moreover, a 3D representation of the spatio-temporal evolution of the conditional probability associated with PM_{10} concentrations has been provided by using *ArcScene* of *ArcGIS*.

Up to now, complete integration between geostatistical tools and *GIS* is not available. Hence, further developments concerning the implementation of scripts in a *GIS* package might be proposed, in order to provide a user-friendly interface for the analysis of spatio-temporal data and their representation.

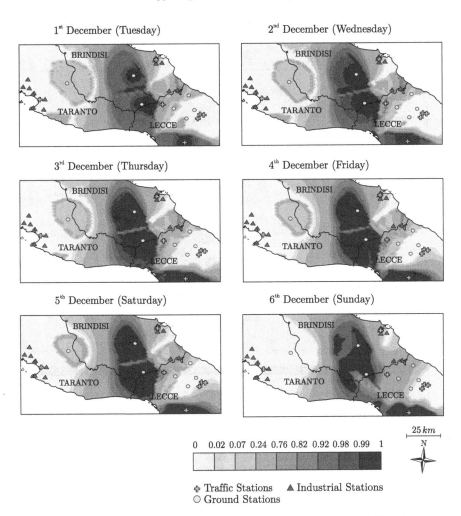

Fig. 4.6 Conditional probability maps of PM$_{10}$ concentrations, for $z_2 = 40.57 \, \mu g/m^3$ (80th percentile), in the Grande Salento, during the period 1–6 December 2009, obtained by using *ArcMap*

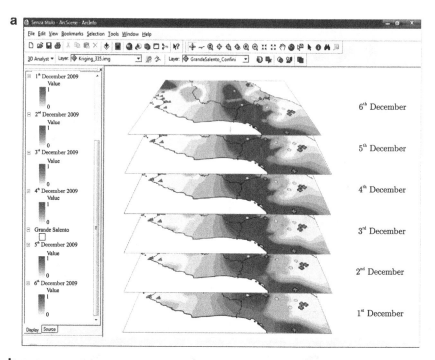

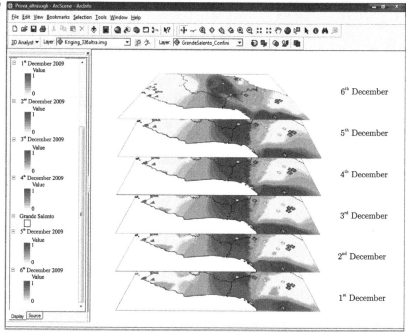

Fig. 4.7 Conditional probability maps of PM_{10} concentrations, in the Grande Salento, during the period 1–6 December 2009, obtained by using *ArcScene*. (**a**) $z_1 = 37.804 \, \mu g/m^3$ (75th percentile). (**b**) $z_2 = 40.57 \, \mu g/m^3$ (80th percentile)

Acknowledgements The authors would like to thank Prof. D. Posa for his helpful suggestions and Prof. S. De Iaco for supporting the research activities involved in this paper by the Project "5 per mille per la ricerca" entitled "Modelli di Interpolazione Stocastica per il Monitoraggio Ambientale: Sviluppi Teorici e Applicativi", University of Salento (2011–2012).

References

1. Anselin, L.: Computing environments for spatial data analysis. J. Geogr. Inf. Syst. **2**(3), 201–220 (2000)
2. Biggeri, A., Bellini, P., Terracini, B.: Metanalisi italiana degli studi sugli effetti a breve termine dell'inquinamento atmosferico. Epidemiol. Prev. **25**(2), Suppl. 1–72 (2001)
3. Bivand, R.S., Gebhardt, A.: Implementing functions for spatial statistical analysis using the R language. J. Geogr. Syst. **2**(3), 307–311 (2000)
4. Boots, B.: Using *GIS* to promote spatial analysis. J. Geogr. Syst. **2**(1), 17–21 (2000)
5. Chen, K., Blong, R., Jacobson, C.: Towards an integrated approach to natural hazard risk assessment using GIS: with reference to bushfires. Environ. Manage. **31**(4), 546–560 (2003)
6. Chilés, J., Delfiner, P.: Geostatistics: Modeling Spatial Uncertainty. Wiley, New York (1999)
7. Cressie, N., Huang, H.: Classes of nonseparable, spatial-temporal stationary covariance functions. J. Am. Stat. Assoc. **94**(448), 1330–1340 (1999)
8. De Iaco, S.: Space-time correlation analysis: a comparative study, J. Appl. Stat. **37**(6), 1027–1041 (2010)
9. De Iaco, S.: A new space-time multivariate approach for environmental data analysis. J. Appl. Stat. **38**(11), 2471–2483 (2010)
10. De Iaco, S., Myers, D.E., Posa, D.: Space-time analysis using a general product-sum model. Stat. Probab. Lett. **52**(1), 21–28 (2001)
11. De Iaco, S., Myers, D.E., Posa, D.: Nonseparable space-time covariance models: some parametric families. Math. Geol. **34**(1), 23–42 (2002)
12. De Iaco, S., Myers, D.E., Posa, D.: On strict positive definiteness of product and product-sum covariance models. J. Stat. Plan. Inf. **141**(3), 1132–1140 (2011)
13. De Iaco, S., Myers, D.E., Posa, D.: Strict positive definiteness of a product of covariance functions. Commun. Stat. Theory Methods **40**(24), 4400–4408 (2011)
14. De Iaco, S., Posa, D.: Predicting spatio-temporal random fields: some computational aspects. Comput. Geosci. **41**, 12–24 (2012)
15. Dimitrakopoulos, R., Luo, X.: Spatiotemporal modeling: covariance and ordinary kriging systems. In: Dimitrakopoulos, R. (ed.) Geostatistics for the Next Century, pp. 88–93. Kluwer, Dordrecht (1994)
16. Diodato, N., Ceccarelli, M.: Multivariate indicator kriging approach using a GIS to classify soil degradation for Mediterranean agricultural lands. Ecol. Indic. **4**(3), 177–187 (2004)
17. Gneiting, T.: Nonseparable, stationary covariance functions for space-time data. J. Am. Stat. Assoc. **97**(458), 590–600 (2002)
18. Goodchild, M.F., Haining, R., Wise, S.: Integrating GIS and spatial data analysis: problems and possibilities. Int. J. Geogr. Syst. **6**(5), 407–423 (1992)
19. Kolovos, A., Christakos, G., Hristopulos, D.T., Serre, M.L.: Methods for generating non-separable spatiotemporal covariance models with potential environmental applications. Adv. Water Resour. **27**(8), 815–830 (2004)

20. Lin, J., Chang, T., Shih, C., Tseng, C.: Factorial and indicator kriging methods using a geographic information system to delineate spatial variation and pollution sources of soil heavy metals. Environ. Geol. **42**(8), 900–909 (2002)
21. Ma, C.: Linear combinations for space-time covariance functions and variograms. IEEE Trans. Signal Process. **53**(3), 489–501 (2005)
22. Poggio, L., Vrscaj, B., Hepperle, E., Schulin, R., Marsan, F.A.: Introducing a method of human health risk evaluation for planning and soil quality management of heavy metal-polluted soils. An example from Grugliasco (Italy). Landsc. Urban Plan. **88**(2–4), 64–72 (2008)
23. Posa, D.: The indicator formalism in spatial data analysis. J. Appl. Stat. **19**(1), 83–101 (1992)
24. Posa, D., De Iaco, S.: Geostatistica. Teoria e Applicazioni. Giappichelli editore, Torino (2009)
25. Spadavecchia, L., Williams, M.: Can spatio-temporal geostatistical methods improve high resolution regionalisation of meteorological variables? Agric. For. Meteorol. **149**(6–7), 1105–1117 (2009)

Chapter 5
Socioeconomic Zoning: Comparing Two Statistical Methods

Silvestro Montrone and Paola Perchinunno

Abstract The aim of this paper is to identify territorial areas and/or population subgroups characterized by situations of deprivation or strong social exclusion through a fuzzy approach that allows the definition of a measure of the degree of belonging to the disadvantaged group. Grouping methods for territorial units are employed for areas with high (or low) intensity of the phenomenon by using clustering methods that permit the aggregation of spatial units that are both contiguous and homogeneous with respect to the phenomenon under study. This work aims to compare two different clustering methods: the first based on the technique of SaTScan [Kuldorff: A spatial scan statistics. Commun. Stat.: Theory Methods **26**, 1481–1496 (1997)] and the other based on the use of Seg-DBSCAN, a modified version of DBSCAN [Ester et al.: A density-based algorithm for discovering clusters in large spatial databases with noise. In: Proceeding of the 2nd International Conference on Knowledge Discovery and Data Mining, pp. 94–99 (1996)]. [The contribution is the result of joint reflections by the authors, with the following contributions attributed to Montrone (Sects. 5.1, 5.3.3 and 5.4) and to Perchinunno (Sects. 5.2, 5.3.1 and 5.3.2).]

Keywords Density-based clustering • Scan statistics • Socioeconomic indicators

S. Montrone • P. Perchinunno (✉)
Dipartimento di Studi Aziendali e Giusprivatistici "Carlo Cecchi", University of Bari "A. Moro",
Via C. Rosalba 53, 70100 Bari, Italy
e-mail: s.montrone@dss.uniba.it; p.perchinunno@dss.uniba.it

S. Montrone and P. Perchinunno (eds.), *Statistical Methods for Spatial Planning and Monitoring*, Contributions to Statistics, DOI 10.1007/978-88-470-2751-0_5, © Springer-Verlag Italia 2013

5.1 Statistical Methods for the Identification of Geographical Clustering

5.1.1 Introduction

The great abundance of data deriving from the use of information technology (IT) and digital mapping (DM), and the frequent use of geographic information systems (GIS), increases the interest in geographical analysis and modelling, to support the heuristic creation of new scientific knowledge. In the field of statistics, such instruments are already numerous; for instance, ordinary data mining (ODM) is used in market investigation in order to assess consumer preferences or consumer profiles.

In the field of spatial analysis, instead, although the consciousness of the need to associate spatial entities with information, geographic data mining (GDM) did not reach the same level of stability of results due to greater computational complexity, in problems like geographic clustering, bordering and modelling, due to the intrinsic characters of spatial data. In fact in ODM some specific issues are: explorative data analysis, thematic mapping, multivariate methods, logistic regression (general linear modeling), and clustering (decision/segment trees).

Sometimes traditional methods are likely to be more useful for ODM studies, due to the difficulty in the management of spatial-based entities. ODM traditionally finds relationships between entities without explaining the connection with spatial location, represented by topographic reference or belonging to a special geometric group. In fact, ODM does not consider topological relationships or applications. In order to answer the above questions, spatial databases (SD) and GIS have been developed starting from the 1970s. SD and GIS, owing to their dynamic-link library oriented to the integration with external software, help to integrate the conceptual and pragmatic dimension of the relationship among spatial entities. An incompletely resolved problem is that GIS, and/or geographic analysis machines (GAM) give only a visualization of problem solution, by heuristics that require very complex computation.

If it is possible to reduce the analyzed problem to a few aspects of the observed phenomenon (such as incidence of diseases and/or crimes in a territory), the cartography becomes a thematic map where areas of interest are joined to each other with spatial contiguity. The result is a zoning based on a spatial (or spatial–temporal) clustering, the conceptual aspects of which deserve a better definition. Knox [17] in his studies on spatial relationship of epidemic phenomena gave a definition of spatial clustering: a spatial cluster is a non-usual collection/aggregation of real or perceived (social, economic) events; it is a collection of spatial, or spatial/temporally delimited events, an ensemble of objects located in contiguous areas.

Referring to a given phenomenon from a statistical point of view, in this case the clustering can be based on the identification of areas where a group of points shows the maximum incidence inside, and at the same time leaves the minimum incidence outside. Such an operation is obtained by locating a circular window of arbitrary radius, by calculating the probability (risk) p_1, inside the circle, or the probability (risk) p_2, outside the circle. The minimum p-value (probability of critical region referring to the test) corresponds to the most important cluster. The identification of a special area can be based on the intensity of a statistical attribute, instead of the number of attribute-characterized elements.

In order to examine the possibility of applying such methods to regeneration programs, it is necessary to introduce a physical reference to urban spaces. In the field of epidemiological studies many research groups have developed different typologies of software; these are all based on the same approach, but usually differ from each other in the shape of the window.

Among the various methods of zoning, there are SaTScan [18] that uses a circular window, FlexScan [27], that uses contiguity to build the window, the upper level scan statistics [26], that underpasses the question of geometric shape of the window including aggregate points and finally AMOEBA (a multidirectional optimal ecotope-based algorithm) [1], that uses a similar approach to SaTScan, without the constraint of a circular window.

5.1.2 SaTScan Method

In this section we review the spatial scan statistics following quite closely the original treatment proposed by Kuldorff in a series of remarkable papers [18].

SaTScan is employed to examine an area of interest with a moving window comparing a smoothing of its internal and external intensity: clusters are formed by aggregating units belonging to contiguous windows with similar intensity. Windows of different sizes are used. The most likely cluster is that with the maximum likelihood, by which we intend the cluster least likely to be due to chance. A p-value is assigned to this cluster.

The description of the SaTScan method which follows assumes that the region being examined can be divided into subareas that share no common points, together with the existence of exactly one subset Z (formed by uniting one or more areas) and two independent Poisson processes defined on Z and Z^c, indicated, respectively, with X_Z and X_Z^c which have the intensity functions:

$$\lambda_Z(x) = p\mu(x) \quad \text{and} \quad \lambda_{Z^c}(x) = q\mu(x), \tag{5.1}$$

where p and q indicate the individual probability of occurrence, respectively, inside and outside the Z zone.

The "background" has an intensity function with a significant digit that varies according to the particular application considered: for example, in epidemiological investigations, it models the spatial distribution of the population at risk.

The null hypothesis $H_0: p = q$ is that the probability of occurrence is not higher within the area considered than it is outside: it is resolved employing the following likelihoods ratios:

$$\Lambda_Z = \frac{\max_{p>q} L(Z, p, q)}{\max_{p=q} L(Z, p, q)} = \frac{\left(\frac{y_Z}{\mu(Z)}\right)^{y_Z} \left(\frac{y_G - y_Z}{\mu(G) - \mu(Z)}\right)^{y_G - y_Z}}{\left(\frac{y_G}{\mu(G)}\right)^{y_G}}, \qquad (5.2)$$

where y_Z and y_G, respectively, represent the number of events observed within the Z zone and the entire region under study, while $\mu(Z)$ and $\mu(G)$ are usually approximated by the consistency of the population "at risk," respectively, within Z and the whole region under investigation.

The advantage of proceeding according to this method is that the most probable cluster is detected by the highest value of the likelihood ratio seen as a function of Z:

$$\Lambda = \max_Z \Lambda_Z, \qquad (5.3)$$

where Z is an appropriate collection of subsets of G, or at least a collection of putative spatial clusters [22].

In addition, when the statistical significance of the cluster area defined on the Z zone that maximizes the likelihood ratio has been evaluated, other secondary clusters may be significant: in almost all cases the clusters considered are those which do not overlap the main cluster [29].

5.1.3 The DBSCAN and Seg-DBSCAN Models

DBSCAN (density-based spatial clustering of application with noise) was the first density-based spatial clustering method proposed [11]. The fundamental idea behind this method is that in order to define a new cluster, or extend an already existing one, it is necessary to establish the neighborhood of a point with a given radius ε that must contain at least a minimum number of points *MinPts*, i.e. the density in the neighborhood is determined by the choice of a distance function for two points p and q, denoted by dist(p, q).

There are two different kinds of points in a clustering: core points and non-core points. A point is a *core point* if it has more than a specified number of points (*MinPts*) within ε. These are points that are at the interior of a cluster. The non-core points in turn are either border or noise points. A *border point* has fewer than *MinPts* within ε, but is in the neighborhood of a core point. A *noise point* is any point that is not a core point or a border point.

So any two core points that are close enough—within a distance ε of one another—are put in the same cluster; any border point that is close enough to a core point is put in the same cluster as the core point; instead noise points are discarded.

The algorithm DBSCAN, which discovers the clusters and the noise in a database according to the above definitions, is based on the fact that a cluster is equivalent to the set of all points in D which are density-reachable from an arbitrary core point in the cluster. The retrieval of density-reachable points is performed by iteratively collecting directly density-reachable points. DBSCAN checks the ε-neighborhood of each point in the database. If the ε-neighborhood $N_\varepsilon(p)$ of a point p has more than $MinPts$ points, a new cluster C containing the objects in $N_\varepsilon(p)$ is created. Then the ε-neighborhood of all points q in C which have not yet been processed is checked. If $N_\varepsilon(p)$ contains more than $MinPts$ points, the neighbors of q which are not already contained in C are added to the cluster and their ε-neighborhood is checked in the next step. This procedure is repeated until no new point can be added to the current cluster C. The greatest advantages of DBSCAN are that it can follow the shape of the clusters and that it requires only one distance function and two input parameters. Their choice is crucial because they determine whether a group is a cluster of points or a simple noise.

So as to limit arbitrariness in the adoption of a value to assign to ε, usually detected by a heuristic procedure, we have developed a new algorithm in this work: *Segmented* DBSCAN (*Seg*-DBSCAN), a modified version of DBSCAN, in which the clusters are aggregated according to multiple levels of value of ε. Levels of ε are defined by fixing a value of $MinPts$ and we analyze the distribution of the maximum radius of the cores that represent groups formed by $MinPts$ points. We then construct a histogram of this distribution and we choose ε in coincidence with the histogram peaks that indicate a proximity of the cores of a cluster. As suggested in literature, we can fix $MinPts$ value to 4, and a number of levels of ε equal to the number of the highest histogram peaks. The final phase of the algorithm involves merging the clusters obtained. The merger of two clusters C_1 and C_2 with different levels of density ε_1 and ε_2 is obtained if:

$$d(C_1, C_2) \leq \max(\varepsilon_1; \varepsilon_2). \tag{5.4}$$

With this modified algorithm, parameter ε is no longer established a priori [21].

5.1.3.1 The Choice of Distance

The function that in these terms links two points A and B of coordinates $A(x_{1A}, x_{2A}, \ldots, x_{KA})$ and $B(x_{1B}, x_{2B}, \ldots, x_{KB})$ is a common function of distance. In certain situations at each point, the n-tuple of coordinates, it may be necessary to assign the intensity to a further phenomenon, that is to assign the weights w_i to the observations in order to change the distance between them. A generic point can, therefore, be represented as $P_i = (x_{1i}, x_{2i}, \ldots, x_{Ki}, w_i)$ with $0 < w_i < 1$.

If A and B are geometrically close to one another with both presenting high values of w, they are hence even more similar to one another, and it is therefore necessary to "reduce" their distance. Conversely, if A and B are geometrically close

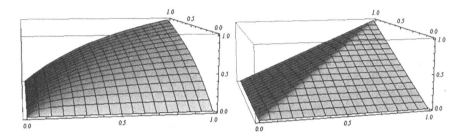

Fig. 5.1 Variation of the function with $t = 1$ or t tending to ∞

to one another but both have low values of w they are not similar, and it is therefore necessary to "enhance" their distance to prevent their fusion into a single core.

According to this logic, the "weighted distance" between the two points is generically obtained through the following formula, given by the ratio between the distance chosen and a mean of order integer $t > 0$:

$$d_{\text{pesata}}(A, B) = \frac{d}{\sqrt[t]{\frac{2}{w_A^{-t} + w_B^{-t}}}}. \tag{5.5}$$

In particular, choosing the Euclidean distance we obtain:

$$d_{\text{pesata}}(A, B) = \frac{\sqrt{\sum_{j=1}^{k} (x_{jA} - x_{jB})^2}}{\sqrt[t]{\frac{2}{w_A^{-t} + w_B^{-t}}}}. \tag{5.6}$$

In this distance the triangle inequality does not hold, so it is a semi-metric, but this restriction does not affect the definitions of density-reachability and density connectivity necessary for DBSCAN algorithm [11]. With this function the distance increases in matching pairs of points with low intensity value, so that they are penalized in the formation of clusters. To vary by t modifies the shape of the function passing from a more rounded to a more angular form, or to a more precise form (Fig. 5.1).

5.1.3.2   Validation Index

The CDbw (Composed Density between and within clusters) is a cluster validity index, proposed by Halkidi and Vazirgiannis [13, 14] that assesses the compactness of clusters and the separation between the clusters generated by an algorithm that takes into account the density distribution between and within the clusters. In particular, the index is based on cohesion of the relative density in terms of intra-cluster density and distance, and it takes account of the different geometric shapes of the clusters and the distribution of noise points between the clusters. The CDbw index is defined as the product of three factors:

$$\text{CDbw}(C) = \text{Cohesion}(C) \cdot \text{Sep}(C) \cdot \text{Compactness}(C), \tag{5.7}$$

where Cohesion(C) means the ratio between the measured density cohesion within the clusters and the density variations observed within the same; Sep(C) means the separation of clusters as measured by the maximum distance between the clusters considering the number of points distributed among the respective clusters and taking account of noise points; finally Compactness(C) means the average relative density within clusters compared to a contraction factor s.

Let **S** be a collection of data and $D = (V_1, \ldots, V_c)$ a partition of the data set **S** in c clusters where each $V_i = (v_{i1}, \ldots, v_{ir})$ is a representative set of r points in cluster C_i. Because the CDbw is well suited to assess results on data sets with clusters of arbitrary shapes, each cluster is not represented by a single point, for example the centroid, but from a collection of r points fairly well scattered within the same cluster, in order to diversify the geometry of the various clusters. We introduce definitions for the construction of the index.

Definition 1 (Closest Representative Points)

Let V_i and V_j be two sets of representative points of clusters C_i and C_j, a point v_{ik} of C_i is said to be the *closest representative points* v_{jl} of C_j, and is denoted by closest_rep$^i(v_{jl})$ if v_{ik} is the representative point of C_i with the minimum distance from v_{jl}, namely

$$d(v_{jl}, v_{ik}) = \min_{v_{ix} \in V_i}\{d(v_{jl}, v_{ix})\},$$

where d is the Euclidean distance.

Therefore the collection:

$$\text{CR}_j^i = \{(v_{ik}, v_{jl})|v_{jl} = \text{closest\_rep}^j(v_{ik})\}. \tag{5.8}$$

Definition 2 (Respective Closest Representative Points)

The set of points is:

$$\text{RCR}_j^i = \{(v_{ik}, v_{jl})|v_{jk} = \text{closest\_rep}^i(v_{jl}); \ v_{jl} = \text{closest\_rep}^j(v_{ik})\} \tag{5.9}$$

is said to be the *respective closest representative points* and is the intersection of the *closest representative* points of C_j with respect to C_i and the *closest representative* points of C_i with respect to C_j that is $\text{RCR}_{ij} = \text{CR}_j^i \cap \text{CR}_i^j$.

The separation is defined in terms of the density of the area between the clusters, by which we mean the area between the *respective closest representative points* of clusters (Fig. 5.2).

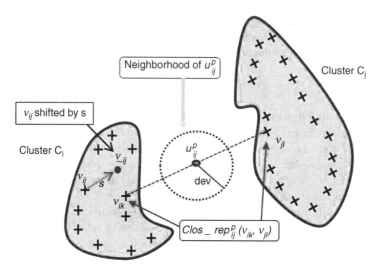

Fig. 5.2 The density within clusters and between clusters for the index CDbw

Definition 3 (Density Between Clusters)

Let $\text{clos\_rep}_{ij}^{p}(v_{ik}, v_{jl})$ be the pth pair of *respective closest representative points* clusters C_i and C_j, so $\text{clos\_rep}_{ij}^{p}(v_{ik}, v_{jl}) \in \text{RCR}_{ij}$, and let u_{ij}^{p} be the middle point of the segment defined by v_{ik} and v_{jl} then the density between the cluster C_i and C_j is defined by:

$$\text{Dens}(C_i, C_j) = \frac{1}{|\text{RCR}_{ij}|} \cdot \sum_{i=1}^{|\text{RCR}_{ij}|} \left(\frac{d(\text{clos\_rep}_{ij}^{p})}{2 \cdot \text{dev}} \cdot \text{cardinality}(u_{ij}^{p}) \right), \quad (5.10)$$

where:

- $d(\text{clos\_rep}_{ij}^{p})$ is the Euclidean distance between the pair of points defined by $\text{clos\_rep}_{ij}^{p}(v_{ik}, v_{jl}) \in \text{RCR}_{ij}$.
- $|\text{RCR}_{ij}|$ is the cardinality of the set RCR_{ij}.
- dev is the average of the difference between the representative points considered.
- $\text{cardinality}(u_{ij}^{p}) = \dfrac{\sum_{l=1}^{n_i+n_j} f(x_l, u_{ij}^{p})}{n_i+n_j}$ with $x_l \in C_i \cup C_j$; n_i and n_j are the cardinality of C_i and C_j, respectively.

In particular, $\text{cardinality}(u_{ij}^{p})$ represents the average number of points C_i and C_j that belong to the neighborhood of u_{ij}^{p}. The function f tests whether or not the point x belongs to the neighborhood of u_{ij}^{p} is:

$$f(x, u_{ij}) = \begin{cases} 1 & \text{se} \quad d(x, u_{ij}) < \text{dev} \quad \text{and} \quad x \neq u_{ij} \\ 0 & \text{otherwise} \end{cases}.$$

The cardinality (u_{ij}^p) is influenced by the definition of neighborhood of a point. An appropriate definition of neighborhood is necessary for each data set, in the specific case of the previous definition the neighborhood considered is a hyper-sphere centered at u_{ij}^p with radius equal to dev.

Considered $C = \{C_i | i = 1, ..., c\}$ a partition of c cluster with $c > 1$ we have the following definitions of density and separation between clusters and density, compactness and cohesion within the cluster.

Definition 4 (Inter-cluster Density)

The maximum distance between C_i and the other clusters in C measure the density of clusters, denoted with *Inter-cluster density*, is defined as:

$$\text{Inter\_dens}(C) = \frac{1}{c} \sum_{i=1}^{c} \max_{\substack{j=1,...,c \\ j \neq i}} \{\text{Dens}(C_i, C_j)\} \tag{5.11}$$

with $c > 1$ and $c \cdot n$.

Definition 5 (Clusters Separation)

The following ratio defines the measure of separation between clusters:

$$\text{Sep}(C) = \frac{\frac{1}{c} \sum_{i=1}^{c} \min_{\substack{j=1,...,c \\ j \neq 1}} \{\text{Dist}(C_i, C_j)\}}{1 + \text{Inter\_dens}(C)} \tag{5.12}$$

with:

- $c > 1$ and $c \cdot n$.
- $\text{Dist}(C_i, C_j) = \frac{1}{|\text{RCR}_{ij}|} \cdot \sum_{i=1}^{|\text{RCR}_{ij}|} d(\text{clos\_rep}_{ij}^p)$, where $|\text{RCR}_{ij}|$ is the cardinality of set RCR_{ij}.

Definition 6 (Relative Intra-cluster Density)

The relative density of clusters compared to a contraction factor, s, is defined as:

$$\text{Intra\_dens}(C, s) = \frac{\text{Dens\_cl}(C, s)}{c \cdot \text{dev}} \tag{5.13}$$

with:

- $c > 1$.
- $\text{Dens\_cl}(C, s) = \frac{1}{r} \sum_{i=1}^{c} \sum_{j=1}^{r} \text{cardinality}(v_{ij})$.
- $\text{cardinality}(v_{ij}) = \frac{\sum_{l=1}^{n_i} f(x_l, v_{ij})}{n_i}$ where n_i is the number of points x_l that belong to the C_i and f is the function as defined earlier.

Definition 7 (Compactness)
The compactness of a partition in terms of density is defined as the average density inside the C partition:

$$\text{Compactness}(C) = \sum_s \frac{\text{Intra\_dens}(C, s)}{n_s}, \qquad (5.14)$$

where n_s is the number of different values considered for s and for which the *intra-cluster density* is calculated, usually s varies in the range $[0, 1; 0, 8]$.

Definition 8 (Intra-density Change)
The variation in density within the clusters is defined with the following ratio:

$$\text{Intra\_Change}(C) = \frac{\sum\limits_{i=1,\ldots,n_s} |\text{Intra\_dens}(C, s_i) - \text{Intra\_dens}(C, s_{i-1})|}{(n_s - 1)}, \qquad (5.15)$$

where n_s is the number of factors of contraction s considered. Significant changes in density within the cluster (*intra-density change*) indicate high-density areas alternate to low-density areas.

Definition 9 (Cohesion)
Cohesion is a measure of the intra-cluster density compared to density variations observed within them is defined through the following ratio:

$$\text{Cohesion}(C) = \frac{\text{Compactness}(C)}{1 + \text{Intra\_Change}(C)}. \qquad (5.16)$$

Definition 10 (Separation Respect to Compactness)
The product of the density between clusters and density within clusters is denoted by:

$$\text{SC}(C) = \text{Sep}(C) \cdot \text{Compactness}(C). \qquad (5.17)$$

Now we have the elements for the definition of the index CDbw: Composed Density between and within clusters.

Definition 11 (CDbw)
The CDbw(C) index has a minimum when all points are considered as clusters whereas it has a local maximum if the data set has a natural tendency to clustering:

$$\text{CDbw}(C) = \text{Cohesion}(C) \cdot \text{SC}(C) \qquad (5.18)$$

with $c > 1$.

The index CDbw(C) can be used:

- to select the best input parameters for each cluster algorithm,
- to select between various cluster algorithms, the algorithm which presents the best results,
- to compare the CDbw(C) index with other indices of validity with respect to the number of clusters identified.

5.1.4 The DENCLUE Model

There are other density-based clustering methods; these include Optics [2] and another clustering algorithm used in large multimedia databases, known as DENCLUE (DENsity-based CLUstEring). This approach is based on the analytical modelling of the overall point density as the sum of influence functions of the data points.

The cluster model employed in the DENCLUE algorithm is based on kernel density estimation. The DENCLUE framework for clustering [15, 16] builds upon Schnell's algorithm. Local maxima of the density estimate are used to define clusters. Data points are assigned to local maxima by hill climbing. The points which are assigned to the same local maximum are included in a single cluster.

The advantages of this approach are:

- It has a solid statistical basis.
- Its clustering properties are good even in data sets that have large amounts of noise.
- It allows arbitrarily shaped clusters in high-dimensional data sets to be described in a compact mathematical way.
- It is significantly faster than existing algorithms.

DENCLUE has the disadvantage that the hill climbing used may make small steps in the beginning unnecessary and, while it does closely approach the maximum, it never completely converges. However, the superiority of this new approach emerges when it is compared to DBSCAN shows. Further developments involve applying a "weighted" DENCLUE, obtained using the intensity of a phenomenon instead of the density kernel.

5.2 Multidimensional Aspects of Socioeconomic Deprivation

5.2.1 Construction of Indicators of Socioeconomic Deprivation

Over recent years, and related in particular to the significant recent international economic crisis, an increasingly worrying rise in poverty levels has been observed both in Italy and in other countries.

Such a phenomenon may be analyzed from an objective perspective (i.e., in relation to the macro and micro-economic causes by which it is determined) or, rather, from a subjective perspective (i.e., taking into consideration the point of view of individuals or families who consider themselves as being in a condition of hardship). Indeed, the individual "perception" of a state of being allows for the identification of measures of poverty levels to a much greater degree than would the assessment of an external observer. For this reason, experts in the field have, in recent years, attempted to overcome the limitations of traditional approaches, focusing instead on a multidimensional approach towards social and economic hardship, equipping themselves with a wide range of indicators on living conditions, whilst simultaneously adopting mathematical tools which allow for a satisfactory investigation of the complexity of the phenomenon under examination [24].

Since the end of the 1970s, numerous studies have been based on a variety of approaches, each of which adopted an attentive definition and conceptualization of the phenomena. Townsend [28] defines poor families as those that "lack the resources for a quality of alimentation, participation in activities and enjoyment of the living conditions which are standard, or at least widely accepted, in the society in which they are living." The reference is, therefore, towards a concept of poverty as relative privation, which takes into account the particular historical, economic, social, geographical, cultural, and institutional context under examination. Within this study, 12 principal dimensions of poverty were identified which are: diet, clothing, housing costs, costs within the household, living conditions, working conditions, health, education, the environment, family activities, recreational activities, and social relations. It may be noted that 3 of the 12 areas considered are connected to housing conditions. The 12 categories described above have been used in many later studies based on the concept of so-called multidimensional poverty, carried out amongst others by Gailly and Hausman [12] and Desai and Shah [9].

With regard to the choice of *poverty index*, there is, therefore, a consideration of various aspects associated with educational levels, with working conditions, and with housing conditions along with the quality of housing. In this case the indices were chosen with the aim of identifying the level of residential poverty and were calculated in order to align elevated levels on the indices with elevated levels of poverty.

A particular index tied to *social difficulty* of the resident population, the *index of lack of progress to high school diploma*, is obtained by elaborating the ratio between the total resident population aged 19 or over who have not achieved a high school diploma and the total resident population of the same age. Such an evaluation presupposes that poverty is in some way tied to levels of schooling, at the very least in cultural terms.

Another important index as a measure of poverty, tied to occupational dynamics, is the *rate of unemployment*, understood as resulting from the ratio between the population aged 15 or over in search of employment with respect to the total labor force of the same age group.

A further measure is the *index of overcrowding*: the ratio between the total number of residents and the size of dwellings occupied by residents. Connected to the phenomena of *housing deprivation* is the evaluation of the *classification of housing status* (in rented accommodation, homeownership, usufruct, or free use). In particular, it is evident that homeownership is an indicator inversely correlated with poverty.

A measure of poverty is, therefore, represented by the *incidence of the number of dwellings occupied by rent-payers* with respect to the total number of dwellings occupied by residents.

Finally, aspects of residential poverty associated with the *availability of functional services* are considered in the analysis, including goods of a certain durability destined for communal use such as the availability of landline telephone or the presence of heating systems. Consistent with the aim of identifying aspects related to poverty, the *incidence of the number of dwellings deprived of services* was calculated for each residence (landline telephone and heating system) with respect to the total number of dwellings occupied by residents.

5.2.2 Statistical Methods for Multidimensional Analysis of Hardship

The different scientific research approaches are consequently directed towards the creation of *multidimensional indicators*, sometimes going beyond dichotomized logic in order to move towards a classification which is "fuzzy" in nature, in which every unit belongs to the category of poor with a range from 1 to 0, where the value 1 means definitely poor, 0 means not poor at all, and the other values in the interval reflect levels of poverty. Classifying populations simply as either *poor* or *non-poor* constitutes an excessive simplification of reality, negating all shades of difference existing between the two extremes of high level well-being and marked material impoverishment. Poverty is certainly not an attribute which can characterize an individual in terms simply of presence or absence, but rather is manifested in a range of differing degrees and shades [8, 25].

The development of *fuzzy theory* stems from the initial work of Zadeh [30], and successively of Dubois and Prade [10] who defined its methodological basis. Fuzzy theory assumes that every unit is associated contemporarily with all identified categories and not univocally with only one, on the basis of ties of differing intensity expressed by the concept of degrees of association. The use of fuzzy methodology in the field of "poverty studies" in Italy dates back only a few years, and is primarily due to the work of Cheli and Lemmi [8] who define their method "Total Fuzzy and Relative" (TFR) on the basis of the previous contribution from Cerioli and Zani [7].

The TFR method defines the measurement of an individual's *degree of member-ship* to the fuzzy totality of the poor, comprised within the interval between 0 (where an individual does not clearly demonstrate membership of the totality of the poor) and 1 (where an individual clearly demonstrates membership of the totality of the poor). In mathematical terms a method of this type involves the construction of a function of membership of "the fuzzy totality of the poor" that is continuous in nature, and "able to provide a measurement of the degree of poverty present within each unit" [8]. Assuming that k indicators of poverty are observed for every family, the function of membership of *i*th family to the fuzzy subset of the poor may be defined thus [7]:

$$f(x_{i.}) = \frac{\sum\limits_{j=1}^{k} g(x_{ij}) \cdot w_j}{\sum\limits_{j=1}^{k} w_j}, \quad i = 1, 2, \ldots, n. \tag{5.19}$$

The w_j are only a *weighting system* [8, 19, 20], as for the generalization of Cerioli and Zani [7], whose specification is given:

$$w_j = \ln\left(1 \Big/ \overline{g(x_j)}\right). \tag{5.20}$$

The weighting operation is fundamental for creating synthetic indexes, by the aggregation of function of membership of each single indicator of poverty. An alternative, by Betti, Cheli, and Lemmi starts from the conjoint use of the coeffi-cient of variation as the first component of the set of weights, with the correlation coefficient as the second component [3–5]. The new set of weights, that is proposed for continuous variables, takes into account two factors, described in the following multiplicative form:

$$w_j = w_j^{(a)} \cdot w_j^{(b)}, \tag{5.21}$$

where:

- $w_j^{(a)} = \cdot\dfrac{\sigma_j}{\mu_j}$ is the coefficient of variation of X_j.

- $w_j^{(b)} = 1 - \dfrac{\sum\limits_{l \neq j} \rho(X_j,X_l)}{\sum\limits_{l=1}^{k} \rho(X_j,X_l)}$ is the complement to one of the ratio between the sum of all correlation coefficients, left out the j array, and the whole sum of correlation coefficients referring to X_j.

Table 5.1 Average of indexes[a] in some Southern Italian metropolitan areas, 2001

	Educational qualifications	Working conditions	Overcrowding	Housing status	Lack of landline telephone	Lack of heating system
Cagliari	47.5	17.6	2.7	18.9	15.2	23.5
Bari	54.9	19.5	3.4	29.4	16.8	10.2
Napoli	58.5	30.5	4.1	44.2	16.9	33.1

Source: Our elaboration of the data from the Population and Housing Census, 2001
[a]*Index 1*—index of lack of progress to high school diploma: ratio between the total number of residents aged 19 or over who have not obtained a high school diploma and the total number of residents of the same age group. *Index 2*—rate of unemployment: the ratio between the total number of residents aged 15 or over who are in search of employment and the workforce of the same age group. *Index 3*—index of overcrowding: the ratio between the total number of residents and size of dwellings occupied by residents. *Index 4*—incidence of the number of dwellings occupied by rent-payers: ratio between the number of dwellings occupied by rent-paying residents and the total number of residents. *Index 5*—incidence of the number of dwellings lacking a landline telephone: ratio between the number of dwellings occupied by residents without a landline telephone and the total number of dwellings occupied by residents. *Index 6*—incidence of the number of dwellings lacking a heating system: ratio between the number of dwellings occupied by residents without a heating system and the total number of dwellings occupied by residents

5.3 A Case Study

5.3.1 Socioeconomic Indices in Some Cities of South Italy

The subject of the case study arises from the necessity to identify geographical areas characterized by situations of poverty in some metropolitan areas in the South of Italy: Cagliari, Bari, and Napoli. With the aim of analyzing the phenomena of poverty on a geographical basis, the work makes use of the data deriving from the Population and Housing Census 2001 carried out by ISTAT; such information allows the geographical analysis in sections according to the census, albeit disadvantaged by the unavailability of the most recent data.

Table 5.1 shows average indexes of poverty for each city. The analysis of social poverty in the above-mentioned cities shows that the percentage of people aged over 19 who have not achieved a *high school diploma* ranges from 47.5 % in Cagliari to 58.5 % in Napoli. As regards to the *unemployment rate* we see very low percentages in Cagliari (17.6 %) and Bari (19.5 %) and higher scores in Napoli (30.5 %). This index highlights the serious social and economic difficulties occurring in some areas. With regard to the *overcrowding rate*, an average of less than

Table 5.2 Composition of absolute values and percentage values of the census sections in some Italian metropolitan areas for conditions of poverty in 2001

Conditions of poverty	Absolute values	Percentage values (%)
Cagliari	1,198	100.0
Well-off	506	42.2
Non-poor	419	35.1
Almost poor	191	15.9
Unquestionably poor	82	6.8
Bari	1,312	100.0
Well-off	691	52.7
Non-poor	349	26.5
Almost poor	170	13.0
Unquestionably poor	102	7.8
Napoli	3,839	100.0
Well-off	1,212	31.6
Non-poor	1,020	26.5
Almost poor	825	21.5
Unquestionably poor	782	20.4

Source: Our elaboration on the data from the Population and Housing Census, 2001

3 inhabitants per 100 m^2 is observed in Cagliari at variance with Napoli where the largest rate of about 4 people on average for every 100 m^2 of housing occurs. As regards the *typology of occupation*, the percentage of the dwellings occupied by renters ranges from a minimum of 18.9 % of residents renting in Cagliari to a maximum of 44.2 % in Napoli. As regards the incidence of *dwellings deprived of landline telephone*, scores vary from a minimum of 15.2 % in Cagliari to a maximum of 16.8 % in Bari and of 16.9 % in Napoli. Even stronger is the difference with respect to *dwellings with no heating system*: higher percentages are observed in Napoli (33.1 %).

In Table 5.2, the TFR measures of poverty, estimated for the total population of the different cities for the year 2001, are classified into four different typologies of poverty in accordance with the resulting fuzzy values: *well-off* (fuzzy value between a minimum of zero and a maximum of 0.25), *non-poor* (between 0.25 and 0.50), *almost poor* (between 0.50 and 0.75), and *unquestionably poor* (between 0.75 and 1).

Considering the set of indicators relating to *socioeconomic deprivation*, 20.4.7 % of the resident population in Napoli is classified by our fuzzy technique in the *unquestionable poverty* class, at variance with much lower percentages in Cagliari (6.8 %) and Bari (7.8 %). Moreover, Fig. 5.3 indicates that there is a clear percentage of *well-off* in Bari (52.7 %) and Cagliari (42.2 %).

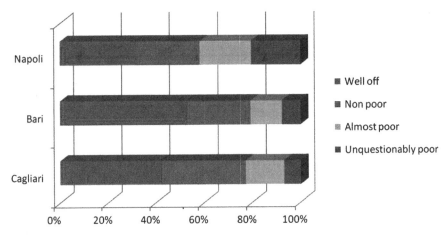

Fig. 5.3 Composition of percentage values of the census sections of Bari, Cagliari, and Napoli for conditions of social and housing deprivation

5.3.2   An Application with SaTScan

In this case, SaTScan operates by locating a circular window of arbitrary radius, and calculating the probability of urban poverty, inside the circle, or the probability of urban poverty, outside the circle, and consequently by optimizing the dimension of the radius [24].

In particular, three cities of the south of Italy have been represented (Napoli, Bari and Cagliari), all with situations of significant social and housing deprivation. In detail, for each city we have identified a different number of clusters, composed of a different number of census sections, where the identification of difficulty is given by the inside average; the higher the average value, the higher is the level of poverty.

A further aspect of interest is given by the p-value, that is the probability of the critical region of the test, where the lower the values shown, the better defined is the cluster.

The interpretation of data analysis shows that the values of the inside averages (included in the interval 0–1) are very high. Optimal values are shown also by p-values. Among the different clusters, some are strongly discriminating, whereas others show a nonsignificant p-value since we either have few cases inside the area or a strong variability is apparent, although around a high value of the inside average (Table 5.3).

The different clusters are shown on a map by different shades of gray ranging from maximum degree of social and housing deprivation (the darkest gray) to minimum degree (the lightest gray), so that the represented reality is immediately understandable (Figs. 5.4, 5.5, and 5.6).

Table 5.3 Description of clusters referring to social and housing deprivation sets (SaTScan method)

Cluster	Number of cases	Inside average	Outside average	p-value
Bari				
1	69	0.67	0.27	0.0010
2	39	0.58	0.28	0.0040
3	480	0.41	0.26	0.0010
Napoli				
1	220	0.65	0.48	0.0010
2	2,226	0.60	0.37	0.0010
Cagliari				
1	2	0.93	0.25	0.3240
2	31	0.77	0.24	0.0010
3	6	0.60	0.25	0.7440
4	505	0.34	0.20	0.0010

Source: Our elaboration of the data from the Population and Housing Census, 2001

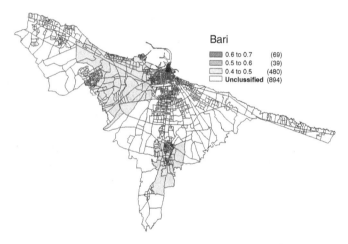

Fig. 5.4 SaTScan model for the identification of Hot Spots of social and housing deprivation in Bari

In detail, two critical areas are observed in *Bari* (Fig. 5.4): the *old center* of the city, called "San Nicola" (with inside average of 0.67), Madonnella and Carrassi (with inside average of 0.58). We have two secondary clusters as well (characterized by a lower average of 0.41), which can be identified with areas in which intense council housing took place during 1960s (Stanic, San Girolamo, Carbonara and Ceglie) [6].

The situation in *Napoli* (Fig. 5.5) is much more critical in that there is a wide area with high average values (ranging from a minimum value of 0.6 to a maximum of 0.7).

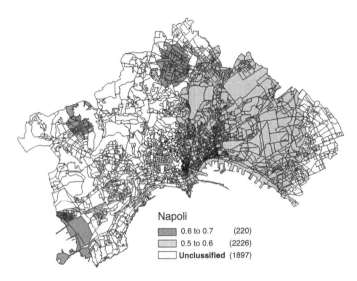

Fig. 5.5 SaTScan model for the identification of Hot Spots of social and housing deprivation in Napoli

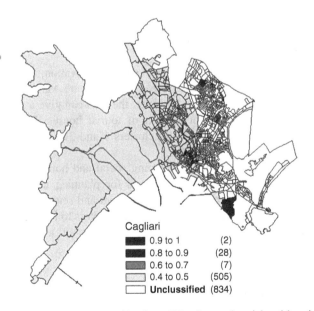

Fig. 5.6 SaTScan model for the identification of Hot Spots of social and housing deprivation in Cagliari

In particular, we can see the presence of areas of maximum deprivation, such as Scampia, Piscinola, Pianura and Bagnoli, where known and serious problems of urban decay occurs. Also, we can see another large area comprising Secondigliano, Miano, S. Pietro Paterno, Poggioreale, Barra, S. Giovanni a Teduccio, Ponticelli, S. Carlo Arena, and S. Lorenzo.

Table 5.4 Description of clusters referring to social and housing deprivation sets (Seg-DBSCAN method)

Cluster	Number of cases	Inside average	CDbw
Bari			0.017181
1	63	0.7–0.8	
2	5	0.6–0.7	
3	219	0.5–0.6	
4	217	0.4–0.5	
5	18	0.3–0.4	
Napoli			0.000083
1	10	0.8–0.9	
2	35	0.7–0.8	
3	566	0.6–0.7	
4	1,431	0.5–0.6	
5	127	0.4–0.5	
Cagliari			0.001638
1	18	0.8–0.9	
2	10	0.6–0.7	
3	141	0.5–0.6	
4	416	0.4–0.5	
5	6	0.3–0.4	
6	5	0.2–0.3	

Source: Our elaboration of the data from the Population and Housing Census, 2001

In *Cagliari* (Fig. 5.6) we have two areas of strong deprivation, where an average of 0.93 is observed. In particular, the areas identified are Manna-Barracca and Levante. In the west side of the city, however, the data can give a misperception, due to the existence of a relevant amount of tourist houses. In this case two secondary clusters are noted (specifically clusters 1 and 3) which present a high p-value (respectively 0.324 and 0.744).

Starting from identified clusters concerning social and housing deprivation, it could be possible to obtain useful indications for planning urban regeneration policies, making decisional process more transparent and scientifically supported. In this way SaTScan offers a possibility for policy makers to localize places urban regeneration interventions should be concentrated, with a methodologically accountable selection of areas [23].

5.3.3 An Application with DBSCAN

The same data on social and housing deprivation were analyzed using the Seg-DBSCAN method of associating the different geographical coordinates with the intensity of fuzzy index of deprivation.

From the analysis of data with the Seg-DBSCAN clusters were obtained for which the average hardship index values are summarized in Table 5.4, also indicating the validation index value of those clusters, as described above (CDbw).

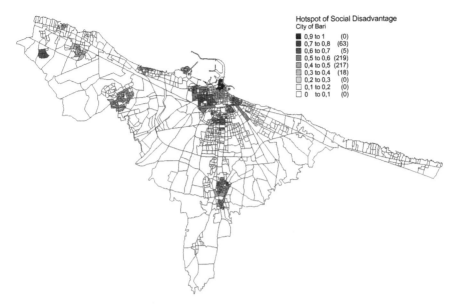

Fig. 5.7 Seg-DBSCAN model for the identification of clusters of socioeconomic deprivation in the city of Bari

The different clusters are shown on a map by a shades of gray ranging from maximum degree of social and housing difficulty (the darkest gray) to minimum degree (the lightest gray) so that the represented reality is immediately understandable (Figs. 5.7, 5.8, and 5.9).

In the city of Bari (Fig. 5.7), five clusters of socioeconomic deprivation may be identified through the use of the Seg-DBSCAN model, consisting of a number of different sections of a census, totalling 525 sections (as compared to 588 with the SaTScan method). This method confirms the presence of areas of socioeconomic hardship in the old city (San Nicola), in neighborhoods adjacent to the city center (such as Madonnella, Libertà, and Carrassi) or those entirely peripheral (San Paolo, Ceglie, and Carbonara). It is particularly evident that the Seg-DBSCAN method is more discriminating than the SaTScan method as it is able to exclude sections from the cluster that do not "effectively" conform to the average values of the cluster, since these areas are subject to redevelopment or not intended for occupancy (schools, public buildings).

The situation in Napoli (Fig. 5.8) reflects what has already been demonstrated with the SaTScan method, with enhanced specification of areas of particular hardship. Specifically, 2,169 sections were revealed as belonging to the clusters as compared with 2,446 sections highlighted with the SaTScan method. The neighborhoods highlighted are Scampia, Secondigliano, Barra, and San Giovanni a Teduccio (which present an internal average of between 0.6 and 0.7) and the industrial area of the city, evidently void of housing, is now excluded having been previously included in the SaTScan mapping. Included within the cluster

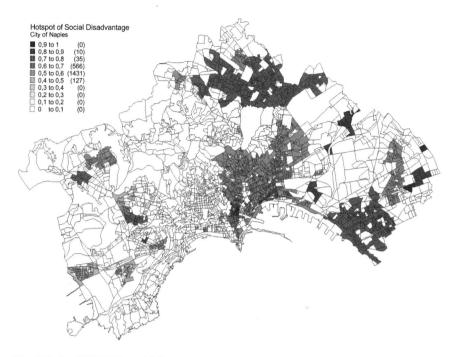

Fig. 5.8 Seg-DBSCAN model for the identification of clusters of socioeconomic deprivation in the city of Napoli

(albeit with average values lower than the previous ones, yet high nonetheless) is the area surrounding the port of Naples, extending towards the San Lorenzo district and the San Carlo Arena.

With regard to the city of Cagliari (Fig. 5.9), 544 census sections are highlighted in the clusters identified through the Seg-DBSCAN method as compared to 596 with the previous method. It would appear evident that the new method is far more discriminating than the previous one in terms of identifying areas more clearly and better-localized in sections of effective hardship while excluding areas not destined for housing. Specifically, the neighborhoods identified are Sant'Elia and C.E.P. with a highly elevated internal average (of between 0.8 and 0.9), San Michele (with an internal average of between 0.6 and 0.7) Sant'Avendrace, Stampace, and Is Mirrionis (with an internal average of between 0.5 and 0.6).

We observe that SaTScan identifies areas formed by contiguous spatial units in which a smoothing of the disadvantaged housing index is performed. This method is effective in identifying areas of high or low intensity and therefore may be a useful indication of areas "at risk" to be monitored. Like the SaTScan method, Seg-DBSCAN identifies areas in which the spatial units meet a criterion of adjacency, but Seg-DBSCAN can exactly identify sections of the city with housing problems, excluding those areas where the phenomenon is absent. For example, in the case of the San Nicola district, the old town of Bari, the SaTScan method

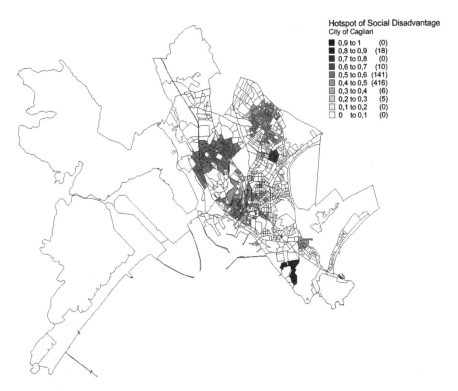

Fig. 5.9 Seg-DBSCAN model for the identification of clusters of socioeconomic deprivation in the city of Cagliari

Fig. 5.10 SaTScan model and Seg-DBSCAN model—zoom of districts of Bari and Napoli

identifies the whole district while the Seg-DBSCAN method identifies the same area of hardship but also analyzes the area in more detail (Fig. 5.10).

The method identifies the particular points with a greater presence of the phenomenon and excludes the points where the phenomenon is not present because of the restoration of historic buildings.

5.4 Conclusions

The ability to describe territorial phenomena by means of an integrated model begins from the construction of multidimensional socioeconomic indicators as a basis for adapting models that can identify zones where there is a risk of social and housing deprivation. When accompanied by careful study, the application of these models to real economic situations has shown that they have a significant capacity to define the areas of deprivation in geographical terms by means of "hot spots," with an evident correspondence between internal characteristics (housing poverty) and external characteristics (social deprivation).

Urban and economic experience tend to suggest that at the same time as the dynamics of the market evolve at speeds that vary greatly between those of the structural characteristics of real estate and those of social context (without rapid and radical changes). It is, in fact, very probable that a representation of the deprivation indicators halfway between censuses would show limited variations of the fuzzy variables in comparison with great changes in the real estate market, as the most recent years show.

The different methodologies proposed here identify areas with a high deprivation index. As we have noted above, the *Seg-DBSCAN* method emerges from the comparison of the two methods as more accurate in identifying the spatial units where housing problems are found. In our future studies, we will seek a statistically sound cluster validity index for spatial data that takes noise points into account and provides the accurate measurement of the *Seg-DBSCAN* method.

The utilization of SaTScan or *Seg-DBSCAN* methodology to identify hot spots of housing deprivation raises certain issues for future social research and urban planning in regeneration areas that are particularly relevant to the European Union policy agenda. The data obtained by the cluster intersection of housing deprivation could provide useful indications for the planning of policies of urban regeneration, making decisional processes more transparent and scientifically valid. Further developments include the application of a "weighed" DENCLUE obtained by using the intensity of a phenomenon instead of the density kernel.

The determination of degraded urban environments indicates how any future general policies in support of housing must take into account the diversities that exist between cities and how urban poverty cannot be considered in the same way in all metropolitan areas.

At a time when public resources for investment are very limited, the first question in seeking town planning and architectonic solutions to the problem of urban regeneration focuses on the identification of areas with the highest urban poverty levels so as to assist political decision-makers in forming their policies in a transparent, carefully planned, and objective manner. It is the opinion of the present authors that the model used in this study is able to provide the data necessary for the accurate identification of such areas.

In conclusion, the model tested here appears to be of value in what the European Union defines as target areas in the context of regional policies for urban regeneration for which there are specific urban plans to be supported with public and private economic and financial resources.

References

1. Aldstadt, J., Getis, A.: Using AMOEBA to create spatial weights matrix and identify spatial clusters. Geogr. Anal. **38**, 327–343 (2006)
2. Ankerst, M., Breunig, M.M., Kriegel, H.-P., Sander, J.: Optics: ordering points to identify the clustering structure. In: Proceedings of the SIGMOD'99, pp. 49–60. ACM, New York (1999)
3. Betti, G., Cheli, B., Lemmi, A.: Studi sulla povertà. Franco Angeli, Milano (2002)
4. Betti, G., Cheli, B.: Poverty dynamics in Great Britain, 1991–1997. A multidimensional, fuzzy and relative approach to analysis. In: Paper for the British Household Panel Survey Research Conference 2001 (BHPS – 2001), Colchester, 5–7 July 2001
5. Betti, G., Verma, V.: Measuring the degree of poverty in a dynamic and comparative context: a multidimensional approach using fuzzy set theory. In: Proceedings of the Sixth Islamic Countries Conference on Statistical Sciences (ICCS-VI), Lahore, pp. 289–301, 27–31 August 1999
6. Campobasso, F., Fanizzi, A., Perchinunno, P.: Homogenous urban poverty clusters within the city of Bari. In: Gervasi, O., Murgante, B., Laganà, A., Taniar, D., Mun, Y., Gavrilova, M.L. (eds.) Computational Science and Its Applications – ICCSA 2008, Part I. LNCS, vol. 5072, pp. 232–244. Springer, Heidelberg (2008)
7. Cerioli, A., Zani, S.: A fuzzy approach to the measurement of poverty. In: Dugum, C., Zenga, M. (eds.) Income and Wealth Distribution, Inequality and Poverty. Springer, Berlin (1990)
8. Cheli, B., Lemmi, A.: Totally fuzzy and relative approach to the multidimensional analysis of poverty. Econ. Notes **24**(1), 115–134 (1995)
9. Desai, M.E., Shah, A.: An econometric approach to the measurement of poverty. Oxford Econ. Paper **40**(3), 505–522 (1988)
10. Dubois, D., Prade, H.: Fuzzy Sets and Systems. Academic, Boston (1980)
11. Ester M., Kriegel H.-P., Sander J., Xu X.: A Density-Based Algorithm for Discovering Clusters in Large Spatial Databases with Noise, Proc. 2nd int. Conf. on Knowledge Discovery and Data Mining (KDD '96), Portland, Oregon, 1996, AAAI Press, 1996
12. Gailly, B., Hausman, P.: Désavantages relatifs à une mesure objective de la pauvreté. In: Sarpellon, G. (ed.) Understanding Poverty. Franco Angeli, Milano (1984)
13. Halkidi, M., Vazirgiannis, M.: Clustering validity assessment: finding the optimal partitioning of a data set. In: Proceedings of IEEE – International Conference on Data Mining (ICDM) Conference, California, pp. 187–194, November 2001
14. Halkidi, M., Batistakis, Y., Vazirgiannis, M.: On clustering validation techniques. J. Intell. Inf. Syst. **17**(2–3), 107–145 (2001)
15. Hinneburg, A., Keim, D.X.: An efficient approach to clustering in large multimedia databases with noise. In: Proceedings of the 4th International Conference on Knowledge Discovery and Datamining (KDD'98), New York, NY, pp. 58–65, September 1998
16. Hinneburg, A., Keim, D.: A general approach to clustering in large databases with noise. Knowl. Inf. Syst. **5**(4), 387–415 (2003)

17. Knox, E.G.: Detection of clusters. In: Elliott, P. (ed.) Methodology of Enquiries into Disease Clustering, pp. 17–20. Small Area Health Statistics Unit, London (1989)
18. Kuldorff, M.: A spatial scan statistics. Commun. Stat. Theory Methods **26**, 1481–1496 (1997)
19. Lemmi, A., Pannuzi, N., Mazzolli, B., Cheli, B., Betti, G.: Misure di povertà multidimensionali e relative: il caso dell'Italia nella prima metà degli anni'90. In: Quintano, C. (ed.) Scritti di Statistica Economica **3**, 263–319 (1997)
20. Lemmi, A., Pannuzi, N.: Fattori demografici della povertà, Continuità e discontinuità nei processi demografici. L'Italia nella transizione demografica. 4 Rubettino, Arcavacata di Rende, pp. 211–228 (1995)
21. Perchinunno, P., Montrone, S., Ligorio, C., L'abbate, S.: Comparing SaTScan and Seg-DBSCAN methods in spatial phenomena. In: Proceedings Spatial Data Methods for Environmental and Ecological Processes, 2nd edn., pp. 115–118. CDP Service Editions, Foggia (2011)
22. Montrone, S., Bilancia, M., Perchinunno, P.: A model-based scan statistics for detecting geographical clustering of disease. In: Gervasi, O., Murgante, B., Laganà, A., Taniar, D., Mun, Y., Gavrilova, M.L. (eds.) Computational Science and Its Applications – ICCSA 2009, Part I. LNCS, vol. 5592, pp. 353–368. Springer, Heidelberg (2009)
23. Montrone, S., Perchinunno, P., Rotondo, F., Torre, C.M., Di Giuro, A.: Identification of hot spots of social and housing difficulty in urban areas: scan statistic for housing market and urban planning policies. In: Murgante, B., Borruso, G., Lapucci, A. (eds.) Geocomputation and Urban Planning, Studies in Computational Intelligence, vol. 176, pp. 57–78. Springer, Heidelberg (2009)
24. Montrone, S., Perchinunno, P., Torre, C.M.: Analysis of positional aspects in the variation of real estate values in an Italian Southern Metropolitan area. In: Taniar, D., Gervasi, O., Murgante, B., Pardede, E., Apduhan, B.O. (eds.) Computational Science and Its Applications, ICCSA 2010. LNCS, vol. 6010, pp. 17–31. Springer, Heidelberg (2010)
25. Montrone, S., Campobasso, F., Perchinunno, P., Fanizzi, A.: An analysis of poverty in Italy through a fuzzy regression model. In: Murgante, B., Gervasi, O., Iglesias, A., Taniar, D., Bernady, O., Apduhan B. (eds.) Computational Science and Its Applications – ICCSA 2011, Part I. LNCS, vol. 6782, pp. 342–355. Springer, Heidelberg (2011)
26. Patil, G.P., Taillie, C.: Upper level set scan statistic for detecting arbitrarily shaped hotspots. Environ. Ecol. Stat. **11**, 183–197 (2004)
27. Takahashi, K., Tango, T.: A flexibly shaped spatial scan statistic for detecting clusters. Int. J. Health Geogr. **4**, 11–13 (2005)
28. Towsend, P.: Poverty in the United Kingdom. Penguin, Harmondsworth (1979)
29. Waller, L.A., Gotway, C.A.: Applied Spatial Statistics for Public Health Data. Wiley, New York (2004)
30. Zadeh, L.A.: Fuzzy sets. Inf. Control **8**(3), 338–353 (1965)

Chapter 6
A Geostatistical Approach to Measure Shrinking Cities: The Case of Taranto

Beniamino Murgante and Francesco Rotondo

Abstract Measuring shrinkage and its effects appears as a fundamental issue in cities' research. Also, shrinkage is a spatial phenomenon defined by data and information based on space dimension relying on a spatial information. The wide use of geo-information is a useful aid to extend common statistic analyses integrating data collected at different levels, comparing data at a municipal level to data referring at census area level (particularly useful for detailed analyses at a neighbourhood scale). Such analyses are particularly suitable for medium and large cities shrinkage analyses, where different neighbourhoods could have different levels of shrinkage and could need distinct strategies to face such phenomenon. Another methodological problem is the interrelation with other spatial units and nearby cities, which can have an influence on urban labour market, economic development, migration flows and housing market. Thereby, the definition of an appropriate regional context is of crucial importance. After an introduction about a comparison between common statistic analyses and geo-statistical methods, with a short literature review, the paper includes an empirical section describing the case of de-industrialized Taranto city, measuring the major indicators of shrinkage, with data referring to census area level, trying to understand if there are shrinking neighbourhoods in the city of Taranto and what is the appropriate regional shrinking context. Then, the paper continues with a section in which the theoretical knowledge is evaluated comparing theory strongholds to main features of shrinkage exemplified by the case of Taranto, trying to contribute to a better understanding of the questions addressed, highlighting the unresolved problems to address some

B. Murgante
Laboratory of Urban and Territorial Systems, University of Basilicata, Via dell'Ateneo Lucano 10, 85100 Potenza, Italy
e-mail: beniamino.murgante@unibas.it

F. Rotondo (✉)
Department of Architecture and Town Planning, Polytechnic of Bari, Via Orabona 4, 70125 Bari, Italy
e-mail: f.rotondo@poliba.it

S. Montrone and P. Perchinunno (eds.), *Statistical Methods for Spatial Planning and Monitoring*, Contributions to Statistics, DOI 10.1007/978-88-470-2751-0_6, © Springer-Verlag Italia 2013

conclusions about still open research challenges. [The contribution is the result of joint reflections by the authors, with the following contributions attributed to Rotondo (Sects. 6.1, 6.2.1 and 6.4) and the others to Murgante.]

Keywords Ecological development • Geostatistics • Socio-economic development • Urban planning strategies

6.1   Introduction

Urban planning in western countries over the past two centuries has developed with the aim of identifying territories involved in governing relentless demographic and economic growth phenomena, often involving overcrowding, traffic issues and housing tensions. Although in the contemporary most populous nations of the world (China, India, Indonesia, Brazil, Russia and Pakistan) cities continue to grow, in western countries there are numerous cities that suffer from obvious demographic and economic contraction phenomena [21]. The causes are numerous[1] and not simple to identify, yet the consequences and phenomena associated with such demographic and economic decline are often similar: increasing numbers of empty properties, stagnation and economic recession, the reduced attraction of the city. A definition of a shrinking city is required in order to better understand such a phenomenon. To date there still exists no widely accepted, unique or shared definition within the international scientific community. Indeed, the phenomenon is described along various lines, often overlapping with distinct concepts such as *urban decline* or *urban decay* [11, 14] coined or previously brought into question [15].

The term *Shrinking Cities* first entered parlance in Germany with the expression "Schrumpfende Städte" during the 1990s and defines, according to the *Shrinking Cities International Network* (SCiRN) research group, densely populated urban areas with a minimum population of 10,000 residents which have suffered a loss of population in the majority of the territory during the previous 2 years and are experiencing economic transformation demonstrating various symptoms of structural crisis [23].

Such shrinking cities have been the subject of studies on urban change in traditional research fields such as demographic and cultural change, urban social geography, suburbanization, deindustrialization and urban regeneration amongst others [30]. It is, therefore, imperative to recognize how such processes may be linked, or contribute towards the phenomenon of the contraction of cities.

Shrinking cities can take on differing characteristics according to the context in which they occur. Studies on the dynamics of growth and decline over longer periods (up to 100 years) assist in an understanding of the phenomenon and, for instance, exemplify why a period of decline may have replaced a previous period of

---

[1] Oswalt [22] and others have attempted to draw up a classification of shrinking cities based on possible causes.

growth, or vice versa. A generalization of the phenomenon remains problematic (as with all social phenomena), since shrinking processes occurring on similar spatial scales or in local contexts with similar characteristics may be substantially different in nature.

A loss of population is often seen as a sign of failure and may imply "losing out" in the national or global competition among cities. Research on the decline of cities across the globe [4, 7, 35] demonstrate population decline as an ancient phenomenon, linked to the natural course of events following demographic processes and economic change.

Research groups studying shrinking cities[2] are, however, highlighting the global dimension of a phenomenon often considered merely as an isolated "incident" in wealthy countries, requiring the investment of public funds and planning for a new period growth in order to resolve the issue. The inexorable decline and the relentless depopulation of cities in the USA, Japan and Western and Eastern Europe have, however, stimulated scientific research towards identifying causes and ways of thinking about urban decline and regeneration.

6.2   Methodological Framework

6.2.1   Measuring Shrinking Cities

The study of cities in contraction provides, in the current social landscape, information essential to the development of regional planning strategies. For this purpose it is important to identify which indicators may be useful in "measuring" urban shrinkage and verify, through such study, the dynamics involved.

The existing literature comparing the evolution of cities in Europe [8] provides a picture of the mode of urban decline (and/or growth) measured to date. Comparative studies are generally based on population indicators [6, 9, 32]. There are, however, numerous examples in which population and economic development do not necessarily go hand in hand. There are, for instance, cases of cities that despite a decreasing population manage to maintain solid economic structure and

---

[2] The first international study on the phenomenon was the shrinking cities project carried out by Kulturstiftung des Bundes in Germany with the support of architect Philipp Oswalt, the Galerie für Zeitgenössische Kunst Leipzig, the Stiftung Bauhaus Dessau and the Archplus magazine (http://www.shrinkingcities.com/, web site visited 24 May 2012). Furthermore, the shrink smart project focuses on how challenges are met by policy and governance systems in various shrinking urban regions (http://www.shrinksmart.de/, web site visited 24 May 2012). Such work is supported by the above-cited Shrinking Cities International Research Network (SCiRN™, http://www.shrinkingcities.org/Home, web site visited 24 May 2012) as well as the Cost Action TU0803: Cities Re-growing Smaller (CIRES, http://www.shrinkingcities.eu, web site visited 24 May 2012). The two authors are members of this last European research group (CIRES).

development and, in others, general demographic stability yet significant problems concerning economic activity and employment.

Population may consequently be considered as an initial premise in urban processes as well as a leading indicator, yet the study of factors in demographic decline provides only a partial view of an issue involving far more complex dynamics.

A set of criteria related to economic and social issues in the context of the development of the population must be considered in order to describe such complexities. Specifically, such indicators regard the housing market, the labour market and economy as well as the environmental and cultural fabric of the case under study.

6.2.1.1 Population Total Evolution, Age, Migration

Three aspects seems to be considered in assessing population development: total evolution, migration and ageing. Population decline may be caused by a natural reduction in the population and/or emigration that, in many cases, leads to an ageing population.

Total Evolution

The natural balance of the population (birth/death rates) provides information on natural changes. Population growth is produced by the increase in births and, therefore, fertility rates and birth rates, essential in identifying current or recent growth in the population.

A decisive role in the evolution of the population is also played by singular events, such as a war or a natural disaster (it is particularly clear observing the so defined age pyramid of a country).

The variation in birth rates in many European countries reflects a change in reproductive behaviour, primarily due to changing political-economic and socio-cultural conditions, often brought about by a new conception of the role of the family and the woman (the second demographic transition). It is also necessary to consider the evolution of women of childbearing age as an expression of the development of future births [13].

Migration

Migration flows and their development, taking into account net migration, represent, together with the natural decline in births, a key factor for the study of urban decline.

In analyzing the phenomenon of migration it is necessary to consider the age of migrants, their place of origin, the characteristics of migrants (gender, age, social status) and causes of migration that may affect such a study in a number of ways.

Immigrants and women of reproductive age produce, for example, a natural decrease in population in their area of origin while, at the same time, a possible population increase in the new place of residence. Furthermore, the age of those migrating and whether or not such movements affect entire families is an essential factor in better understanding population dynamics. Indeed, in this case such a factor does not significantly affect the percentage of births. Migrants aged between 18 and 35 years old play a particularly decisive role [33] during phases of the creation of a family, in terms of both employment and education.

Birth rates and, above all, the abandonment of a city by the young lead to changes in population structure in relation to the age of those remaining (in particular, the selectivity of migration linked to ageing is largely responsible for such phenomenon).

To better understand the dynamics of the city motives for migration must be understood; the principal grounds for such movement is the search for jobs or on educational grounds resulting in long-distance migration [33]. This is accompanied by the desire to improve the quality of life in an environmental/aesthetic sense, health or the social environment, cases which may not necessarily see long distance migration.

The separation of internal (suburbanization) and external migration can also be significant as regards the employment conditions of migrants (those moving to the suburbs and not necessarily losing their job in the city). Migration due to economic downturn tends to produce the largest occupational impact on a city, together with demographic change [13].

Migration in search of employment by those willing to relocate over long distances is, furthermore, decisive in terms of development patterns in the local area of origin that cannot compensate for the draw of employment opportunities elsewhere.

The skills of the migrant population should also be considered as this phenomenon largely affects those with high levels of education and highly qualified personnel. This would also appear to be the case for migration over short distances.

Short distance migration mainly seeks to optimize lifestyles and, in analogy with the theory of the life cycle, such migration generally involves families and the elderly [13]. This is reflected in changes to age structures in the city and surrounding areas whilst not, however, significantly affecting the labour market [10]. The impact of this type of migration may, however, be identified with the increase of the total population in surrounding areas. Such dynamics characterize established phenomena of suburbanization in towns and cities. Migration and natural population growth can occur over similar time-scales yet demonstrating completely different dynamics.

Migration may suddenly and radically change while the fertility rate is characterized by set time-scales (a set time between birth and childbearing age). In contrast to the continuous loss of population, singular events trigger a rapid decline in the birth rate or increase migration in the short to medium term and, possibly, even the long term. A decline in population may result from individual events (episodic population) as, for example, natural disasters, war and political transformation processes such as those of 1989/1990 in Eastern Europe following the disintegration of the Soviet Union.

Ageing

Ageing increases community dependence on younger age groups resulting in a potential decrease in the per capita wealth produced [13]. The most widespread indicator in measuring such a phenomenon is the *dependency ratio*, commonly employed in reports on world population trends by the Department of Economic and Social Affairs of the United Nations.[3]

6.2.1.2 Social Indicators

Structural economic change (such as the loss of the industrial base in western countries) can provoke a loss of jobs and population, when combined with demographic changes, in a highly problematic vicious circle. Socio-demographic change plays a decisive role due to the concurrence of several demographic and economic causes and factors.

As mentioned above, migration is highly selective (mostly affecting the young and qualified), with consequences in the areas concerned; those who migrate no longer play a part in the local labour market. This favours the approach of supply towards demand for both the qualified (job vacancies) and unskilled (LTU) labour market, as well as a quantitative increase. The economy is also affected due to the remaining population (the poor, the elderly, trainees and immigrants) being frequently discriminated from the labour market (low-skilled workers, unemployment) [32].

The income of the population indicates the state of economic well-being and a reduction in wages will result in lower family incomes and an increase in economic disparity. In addition, growing unemployment could cause an increase in the number of people living below the poverty line.

The situation of the young is particularly critical. Without access to jobs their entry into the labour market is consequently delayed, accompanied by a dwindling in the creation of new families who may struggle to identify employment opportunities. Similarly, the decline in the attraction of a city as a place for education may provoke a decrease in the number of students enrolled in compulsory education and, therefore, the closure of schools (preschool and school facilities) resulting in reduced investment in education. Students are, furthermore, essential for the development of a city as they require accommodation, food, supplies and educational facilities and may increase the skill levels of the population. The

---

[3] The Department web site (http://esa.un.org/unpd/wpp/Excel-Data/population.htm) reports old-age dependency ratios for all nations in the world.

presence of low-skilled citizens in the labour market could result as disadvantageous for the economy of a city.

Levels of education and, in particular, labour, schools and recreational facilities are therefore crucial in the educational and cultural life of a shrinking city.

6.2.1.3 Economic Factors

Population decline reveals a strong correlation and interrelation with economic development. Indeed, economic analysis can be described by structural economic change in cities.

The central problem is not the transformation of the economic base but, rather, its extensive erosion. Traditional production may not, in this case, be replaced by modern services or other branches of the service sector [33] that have a corresponding effect on social structure and space. Structural economic crisis sees the economy as a whole shrink due to changing economic conditions, possibly followed by a long period of contraction [34]. Such circulation in economic cycles is characteristic of the theory of long waves and the duration of product life cycle. This can have global economic consequences yet may simply relate to specific areas, such as the dispersion of textile and mining towns in northern England [12] or, indeed, the present study of the crisis in the steel industry in Italy and, in particular, in Taranto [26].

The above examples relate to "single product" economic areas whose main characteristic is the long period of decline in demand resulting in a crisis that tends to spread to other companies, industries or economic sectors in the same region or city [12].

In addition to long-term processes of economic restructuring due to globalization and the destruction of the industrial base, especially in established industrial cities, economic change may also be recorded over a relatively short period. The processes of economic restructuring (globalization and de-industrialization) differ according to both regional contexts and city size [12].

The contentious parameter of gross domestic product (GDP) is often still applied in the absence of viable alternatives when describing changes in the economy (this fundamental question has been debated many times and by many authors, but yet no results have been reached in the real policies to change this insufficient indicator).

Employment dynamics provide a quantitative indication of the effects of economic changes on the population and, therefore, on society.

The study of unemployment must be analyzed in conjunction with the development or loss of jobs while taking into account the possibility of immigration movements. The indicator of unemployment is, in itself, incapable of considering this effect in its entirety. From the combination of the two indicators it is, however, possible to draw conclusions on the state of the labour market.

6.2.1.4 Housing Market Indexes

Population dynamics (whether natural or due to migration) has a significant impact on real estate market and housing development. A declining population may reflect the decrease in the number of families and it may determine a declining house's request. Socio-cultural variations in behaviour and lifestyle may also lead to changes in family structure away from a traditional multigenerational family. Indeed, many societies witness an ever-increasing number of mononuclear families. This change is reinforced by selective emigration (young singles or couples) and increased life expectancy (greater proportion of the elderly living alone). Nevertheless, the overall number of families may not, in fact, decrease, due to the growing number of smaller households (single person families). In Italy, the number of family's components has been declining from 3,3 persons for each family in the 1971 to 2,2 in 2010.[4] The family structure is changing (in terms of a transformation in household structure) and with it the needs of the family thus generating changes in demand with a consequent affect on property values (land and house prices). Contraction in cities may, moreover, involve a decrease in construction; in addition to a shift in price levels (rent, land and housing prices), financial problems can be caused by the presence of numerous vacant properties [25].

It should, however, be noted that the housing market is closely linked to the functions of a city (as, for example, the functions of a regional capital, tourist city, etc.). In order to study *housing indicators*, it is therefore necessary to examine the nature of the real estate market in question including, for example, the type and segmentation of the request or the attractiveness of the city (urban added value).

6.2.1.5 Environmental Factors

Environmental aspects could play a significant role in a shrinking city. Lower population densities may constitute the better use of the environment by citizens, reducing the environmental impact related to urbanization, thus improving the quality of life of residents.

A crucial aspect in ecological issues is that of time. Ecological changes may occur over the relatively short or medium term, while their consequences are generally medium to long term in nature.

In the case of the city of Taranto the long presence of the steel industry and the high pollution rates produced were the significant factors in the phenomena of suburbanization.

Some young couples have chosen to live near Taranto in a smaller town but with a better environment [27].

---

[4] Data derived from www.istat.it, web site visited 24 May 2012.

6.2.1.6   Methodological Issues

Initial difficulty lies in the fact that population decline takes on distinct characteristics from city to city; the second issue concerns the choice of indicators that may best describe the phenomenon.

Such indicators provide a starting point for the analysis of shrinking cities; yet, it is necessary to identify the relationships and interrelationships between such parameters.

Economic and demographic dynamics cannot, for example, be treated separately since they are often interrelated. The abandonment of cities could coincide with the reduction of job opportunities in highly qualified companies. Unemployment may, for example, result as the consequence of declining business, leading to emigration and increasing the dependency ratio, thus causing an aging population.

Having identified indicators of contraction study should be focused on their evolution over time in order to better understand dynamics in the medium to long term. The choice of the time intervals becomes critical as they affect results, considering that neither growth nor contractions are linear processes. Data availability is also crucial for comparative analysis and is dependent on both times and scales.

A further methodological problem occurs when comparisons are performed between countries as a result of different detection methods. The question of comparative indicators of a small European range is highly complex as even countries outside the European Union do not all possess equivalent indicators and several new Member States have not yet adopted European standards.

At the end, the majority of geo-statistical works about shrinking use data at a municipal level to measure shrinking phenomena, but, in our opinion, it seems insufficient to evaluate medium and big shrinking cities phenomena, because very often they are linked to suburbanization or changes in the neighbourhoods dimensions, caused by urban degradation or subsequent regeneration process. That is why, in the next paragraphs, we have tried to identify spatial concentration of urban shrinking in micro census zones inside a medium city such as Taranto in the Apulia Region, already evaluated as a shrinking city [26] and the nearest municipalities probably affected by phenomena of suburbanisation. These micro census zones named "sezioni" in the Italian Census survey.[5]

6.2.2   Spatial Statistical Techniques Applied to Shrinkage Phenomena

The identification of spatial concentration of urban shrinking has been achieved adopting spatial autocorrelation techniques.

---

[5] Istat, Censimento generale della popolazione, 1991, 2001, available on www.istat.it visited 24 May 2012.

The concept of spatial autocorrelation directly derives from the first law of geography by Tobler [31]: "All Things Are Related, But Nearby Things Are More Related Than Distant Things."

Analyzing the spatial dimension of statistical data it can be noticed that they are not mutually independent, values of a spatial unit phenomenon tend to influence values of contiguous spatial elements, and a certain degree of interdependency occurs in all directions due to the interaction with other neighbouring elements.

Considering two elements i and j in a set of n objects, traditional approaches to data analyses take into account the degree of similarity of attributes i and j, at the same time spatial autocorrelation considers the degree of similarity of location i and j, also. Spatial autocorrelation does not analyze a phenomenon along a single direction, but it considers all possible relationships of an element with its surrounding spatial units in all directions. This is important to understand whether a phenomenon is isolated or it has a good level of interaction with its surrounding elements.

The concept of contiguity can be defined as a usually symmetrical generalized matrix of W weight, representing the pattern of connections or ties and their intensity [5], where w_i weights denote the effect of spatial unit j on unit i. Generally a dichotomic contiguity matrix has been adopted where $w_{ij} = 1$ if the i area touches the boundary of j area, and $w_{ij} = 0$ if otherwise. Distances of centroids of spatial elements, square of distances, inverse of distance, etc. can be considered instead of value 1. In the case of point data, it is also possible to define a critical distance beyond which two events will never be adjacent. If the elements are included within this distance, i and j are contiguous, and w_{ij} will be equal to 1; otherwise, w_{ij} will be equal to 0.

Both global and local autocorrelation indexes have been adopted in analyzing urban shrinkage phenomena. Global indicators of autocorrelation are useful to identify if a spatial interdependence exists or not without describing where the phenomenon is concentrated. Local indexes of autocorrelation define where highest or lowest levels of autocorrelation are located.

In this study, Moran Index (I), corresponding Moran scatter plots and Local Indicator of Spatial Association (LISA) have been calculated. Moran I provides an overall measure of spatial autocorrelation [16], Moran scatter plot [1] allows to achieve a graphic representation of spatial relationships and enables us to investigate possible local agglomerations, whilst LISA allows us to take into account local effects of the phenomenon [2, 3].

6.2.2.1 Moran's I Statistic

Moran Index (1948) can be formalized as follows:

$$I = \frac{n}{s_0} = \frac{\sum\limits_{i=1}^{n}\sum\limits_{j=1}^{n}(x_i - \bar{x})(x_j - \bar{x})w_{ij}}{\sum\limits_{i=1}^{n}(x_i - \bar{x})^2} \tag{6.1}$$

where:

- x_i is the variable observed in n spatial partitions and $\bar{x}$ is variable average.
- W_{ij} is the generic element of contiguity matrix.
- $S_0 = \sum_{i=1}^{n} w_{ij}$ is the sum of all matrix elements defined as contiguous according to the distance between points-event. In the case of spatial contiguity matrix, the sum is equal to the number of non-null links.

Since the expression of spatial dependence refers to the connection between nearest units, prior of autocorrelation concept, there is the problem of expressing the degree of proximity of areas by defining the concept of spatial contiguity [20]. Index values may fall outside the range $-1; +1$. Moreover, in case of no autocorrelation the value is not 0 but it is $-1/(n-1)$. So if:

- $I < -1/(n-1) =$ negative autocorrelation,
- $I = -1/(n-1) =$ no autocorrelation,
- $I > -1/(n-1) =$ positive autocorrelation.

A positive and significant value of such statistic indicates that similar values of the variable analyzed tend to characterize contiguous localized areas. In contrast, a significant negative value of Moran-I indicates the presence of dissimilar values of the variable in contiguous areas. The significance of the index does not imply absence of autocorrelation, i.e. the presence of a random distribution of the variable in space.

Moran's index, however, does not allow to evaluate if the general positive spatial dependence corresponds to territorial clusters of regions with high or low level of specialization. It is also possible that the degree of spatial dependence between various different groups within the sample is characterized by the existence of a few clusters, located in specific parts of the study region. Considering these limitations, Moran scatter plot has been adopted.

6.2.2.2 Moran Scatter Plot

GEODA software [1] allows to build Moran scatter plot together with the calculation of Moran's I. The graph represents the distribution of the statistical unit of analysis. Moran scatter plot shows the horizontal axis in the normalized variable x, and on the normalized ordinate spatial delay of that variable (W_x).

The first and third quadrants represent areas of values with positive correlations (high–high, low–low) while the second and fourth quadrants represent areas in negative correlation.

However, Moran scatter plot gives no information on the significance of spatial clusters. The significance of the spatial correlation measured through Moran's I and Moran scatter plot is highly dependent on the extent of the study area.

In case of a large territory the measure does not take into account the presence of heterogeneous patterns of spatial diffusion. Moran's I cannot identify outliers present in the considered statistical distribution. LISA allows to consider local effects related to the phenomenon.

6.2.2.3 Local Indicators of Spatial Association

The currently most popular index of local autocorrelation is the so-called LISA [2, 3]. This index can be locally interpreted as an equivalent index of Moran. The sum of all local indices is proportional to the value of Moran one.

The index is calculated as follows:

$$I_j = \frac{\sum_j w_{ij}(y_i - y)(y_j - y)}{\sum_i (y_i - y)^2} \tag{6.2}$$

with: $\sum_i Ii = \gamma \cdot I$.

For each location, it allows to assess the similarity of each observation with its surroundings. Five scenarios emerge:

- Locations with high values of the phenomenon and high level of similarity with its surroundings (high–high), defined as HOT SPOTS.
- Locations with low values of the phenomenon and high level of similarity with its surroundings (low–low), defined as COLD SPOTS.
- Locations with high values of the phenomenon and low level of similarity with its surroundings (high–low), defined as potential "Spatial Outliers".
- Locations with low values of the phenomenon and low level of similarity with its surroundings (low–high), defined as potential "Spatial Outliers".
- Locations devoid of significant autocorrelations.

LISA can effectively bind a measure of the degree of spatial association relative to its surroundings to each territorial unit, allowing to highlight the type of spatial concentration for the detection of spatial clusters.

6.3 The Case Study

The application has been developed in southern Italy, more particularly Taranto and its surrounding municipalities have been considered in quantifying the shrinkage (Fig. 6.1).

Taranto has 191.810 inhabitants distributed over 209.64 km^2, it was one of the main centres of Magna Grecia, it has the second Italian trading port for freight traffic, mainly connected with Asia and it has important industries in the fields of iron, steel and oil refinery. The localization of these activities generated a great dwelling demand, complied with the construction of very intensive

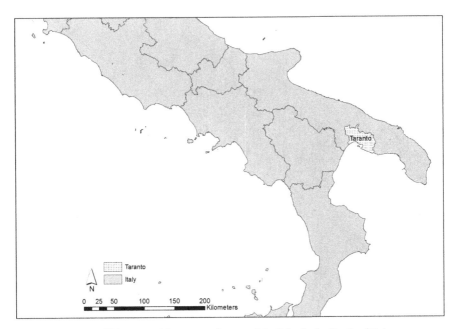

Fig. 6.1 The city of Taranto and its surrounding municipalities in the South of Italy

neighbourhoods. Such a disordered growth realized without master plan, produced urbanization in areas largely disconnected and without continuity. These activities produced a lot of health and environmental problems. Taranto is one of the most polluted cities in Western Europe due to industrial emissions. From this description it can be easily imagined that Taranto could be a very interesting case study in measuring shrinkage.

6.3.1   A Demographically Declining Territory: Taranto City

A brief analysis of the major demographic, economic and social trends at a municipal level highlights possible relations with the dynamics of urban evolution, leading to a clear understanding of the high level of shrinkage in the city. The analysis of the last available census (2008) by the Italian National Statistical Institute (ISTAT[6]) revealed the demographic dynamics (population trends, Net migration, natural balance) of the city are in a constant negative trend, from 1981 to the present. Indeed, the population of Taranto fell from 244,101 in 1981 to

---

[6] All the statistics cited are taken from www.istat.it.

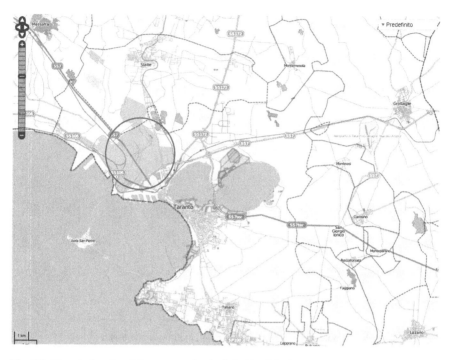

Fig. 6.2 Taranto the city of the two sea. On the *left side* of the image the whole area of the city is dedicated to the ILVA steel plant. (*Source*: OpenStreetMap.)

194,021 in 2008. Net migration is consistently negative from 2002 to 2008 and the city loses, on average, 1,123 inhabitants per year.

The natural balance (live births/deaths) has been constantly negative with the exception of the years 2004 and 2008. In these two years the natural balance is, in any case, significantly below the Net migration as previously shown, with an average of −1,123 inhabitants per year with the overall balance therefore remaining negative. Globalization and the subsequent de-industrialization of European economies is a major cause of urban shrinkage [4, 12, 21]. The relationship between the cycles of the capitalist economy, the life cycles of the city and the effects of globalization on cities and urban regions has been the subject of much study, by authors such as Saskia Sassen.

In the case of Taranto the unemployment rate for the entire Province is 4 % higher than the Apulia Region (18 % with respect to 14.7 % for the Apulia Region as a whole). Indeed, the percentage of those employed in the city of Taranto is 78 % as compared with 80 % in the Province and the Region. Moreover, the total percentage of those in search of employment in the city of Taranto is 22 %, as compared with 20 % in the Province and the Region. Taranto shows a low level of employment compared within its local region .

In the city of Taranto, as in other European cities, the service sector accounts for the largest number of those employed, although traditional industry still accounts for 25 % of total employees (13,767 employees in an industrial sector with 55,174 employees in total). The largest steel plant in Europe is located in Taranto which still employs around 13,346 workers, accounting for almost 100 % of employment within the manufacturing sector in the city. The plant was founded during the 1960s as a state-owned company, under the name "Italsider" in line with fashionable economic and industrial theories of the day regarding large industrial poles (Fig. 6.2). In 1995, after a long crisis in terms of both turnover and employment, the company was sold to the Riva Group (www.rivagroup.com), a major Italian industrial group that operates in the steel industry. Employees in the iron and steel industry are still today predominantly located within younger age groups (21–30 years) with only 23 % exceeding 40 years of age. The steel industry is still, therefore, of fundamental importance to the local labour market of the city of Taranto and its neighbouring municipalities, a plant which would seem extremely difficult to decommission given the long period of economic crisis engulfing western countries [26].

6.3.2 Choosing Indicators for Taranto

The following variables, previously tested in another study [19], have been considered in order to evaluate shrinkage phenomena:

- Dependency ratio is considered as an indicator of economic and social significance. The numerator is composed of people who, because of age, cannot be considered economically independent (youth and elderly), and the denominator of the population older than 15 and younger than 64, who should provide for their livelihood. This index is important in analyzing urban shrinkage because economically active population highlights a degree of vitality in city. While a low level of economically independent population coupled with low birth rate denotes a large presence of hold population.
- Foreign population per 100 residents. Normally foreign number is considered as capability attractiveness, but in southern Italy, where concealed labour rate is 22.8 % and unemployed rate is 20 %, immigration phenomena can be considered a threat and not an opportunity [17, 29].
- Unemployment rate undoubtedly is an important indicator of economic urban decay, which prospects future migration scenarios.
- People living in rented flats. In Italy dwellings ownership rate is more than 80 %; consequently, if resident population lives in rented flats this implies a low-income. In Italy the percentage of tax evasion is high; consequently dwellings ownership is an indicator of economic robustness.
- Per cent of population which had never been to school or dropped out school without successfully completing primary school programs: these indicator denotes the poor quality of social services and social programmes in education.

Table 6.1 Moran Index at 1991 and 2001

Indicator	Moran's I (1991)	Moran's I (2001)
Dependency ratio	0,4860	0,5580
Unemployment rate	0,3808	0,3427
People living in rented flats	0,1410	0,2572
Foreign population per 100 residents	0,027	0,3452
Population which had never been to school or dropped out school without successfully completing primary school program	0,2882	0,3117
Number of people per room in flats occupied by residents	0,1059	0,1515

- Number of people per room in flats occupied by residents. People living in very crowded flats is an underdevelopment indicator because a household can have a flat without respecting the minimum standards, 33 m^3 for inhabitant.

6.3.3 Spatial Distribution of Urban Shrinking

Spatial data have been considered at buildings scale and polygons have been converted in points. Attributes have been associated with such data using census data. In particular, census data of 1991 and 2001 have been adopted. As previously explained, Moran Index is a global indicator of autocorrelation, able to detect a tendency in the whole study area, without precisely defining where the phenomenon is more concentrated. Despite Moran Index lacks in giving a detailed spatial location, it is important for a general analysis of the phenomenon.

Table 6.1 highlights that autocorrelation occurs, in most cases in significant way, for great part of the considered variables. The only Moran Index value close to zero is the Foreign population in 1991, which represents the beginning period of migration phenomenon. Despite this low value, it is important to notice a large increase in transition from 1991 to 2001, where Moran Index reaches a medium level of autocorrelation. The comparison of this index between two different dates allows to assess the phenomenon trend over time.

Data concerning phenomenon concentration have to be compared with the decrease in total population; it means that despite the presence of less people, occurrences of events are more clustered. Considering that the population decrease between 1991 and 2001 in Taranto is 7.2 % and that, at the same time, this reduction arises throughout the whole province, it means that several phenomena are more concentrated in few parts of the study area, increasing the difference in urban quality.

Moran scatter plots at both dates and for all six variables (Figs. 6.3 and 6.4) have been calculated considering standardized variables as abscissa, and spatial weighted standardized variables as ordinate. In the graph, Moran Index corresponds to the direction coefficient of linear regression which represents the scatter plot.

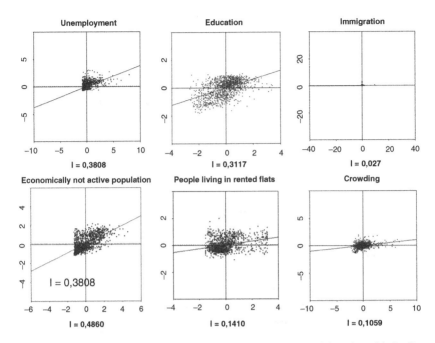

Fig. 6.3 Moran scatter plot for the six variables in 1991. (*Source*: our elaboration with GeoDa on ISTAT data.)

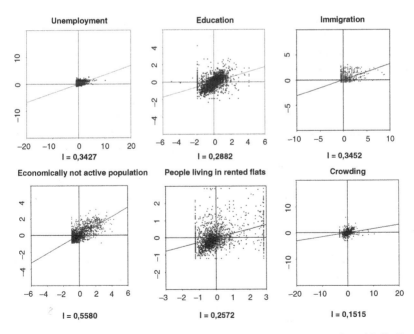

Fig. 6.4 Moran scatter plot for the six variables in 2001. (*Source*: our elaboration with GeoDa on ISTAT data.)

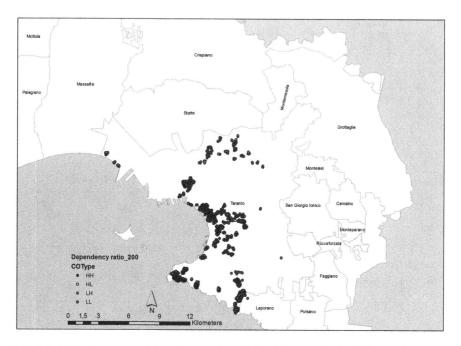

Fig. 6.5 LISA cluster map of dependency ratio, with fixed distance band of 200 m

Spatial autocorrelation has been classified according to Moran scatter plot quadrants. Positive autocorrelation corresponds to spatial clusters upper right (high–high) and lower left (low–low) quadrants. Lower right (high–low) and upper left (low–high) can be classified as spatial outliers.

Figures 6.3 and 6.4 show that the slope of Moran Index is concentrated within the first and fourth quadrants; consequently spatial autocorrelation is positive.

Despite global spatial autocorrelation analysis generates just a value which summarizes the whole study area, the significance of results encourages to apply local autocorrelation index. In many geographical applications, it is highly possible that similar values are located very close to each other. LISA Index allows to discover where the phenomenon is more clustered. After interpreting the current tendency using Moran Index for 1991 and 2001, it is important to understand the place where adopted variables have comparable values. In the study case the attention will be completely paid to 2001 data.

In results achieved adopting LISA, only "hot and cold spots" and potential "Spatial Outliers" have been visualized on the maps while elements without significant autocorrelations have not been showed in order to ensure a clearer visualization. As previously explained, the central aspect characterizing the spatial component in autocorrelation is the weight matrix W.

In the case of point data, the only possible elements of weight matrix can be calculated adopting a fixed distance band. If the spatial unit, which represents buildings, is included within this distance, elements are considered contiguous.

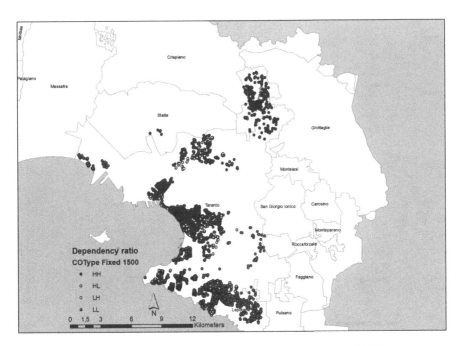

Fig. 6.6 LISA cluster map of dependency ratio, with fixed distance band of 1,500 m

Therefore the choice of such a distance is fundamental in order to achieve good results [18].

As an example, considering Figs. 6.5 and 6.6 the same variable, dependency ratio collects very different results of LISA index adopting a distance of 200 and 1,500 m, respectively.

In this case a distance of 200 m (Fig. 6.5) includes few buildings in a neighbour; consequently, it is not enough to compare the similarity of a variable of contiguous elements. In Fig. 6.6 the dependency ratio with a distance of 1,500 m collects good results, showing that the phenomenon is mainly concentrated in the old part of Taranto city, in the Talsano, Salinella, Tamburi and Paolo VI neighbourhood and in Montemesola municipality. These results highlight a concentration in the direction between Taranto and Martina Franca, the second largest and more populated city of the Province after Taranto.

Obviously, length of the distance is related to study area dimension. In the case of a small municipality, 200 m could be enough.

Analyzing the variable "number of people per room in flats occupied by residents", which can be considered as a crowding index (Fig. 6.7), it is very clear that officially this is not a problem for the area. Nevertheless, the large amount of non-regularized foreign people misrepresents the index in a conspicuous way. In Italy, a strong increase in foreign residence permits has been registered in 2002, following the approval of a law by Italian parliament, concerning immigration discipline and rules.

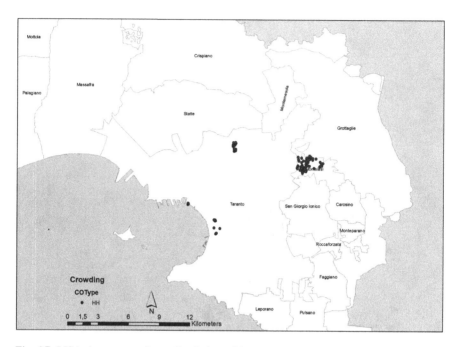

Fig. 6.7 LISA cluster map of crowding index, with fixed distance band of 1,500 m

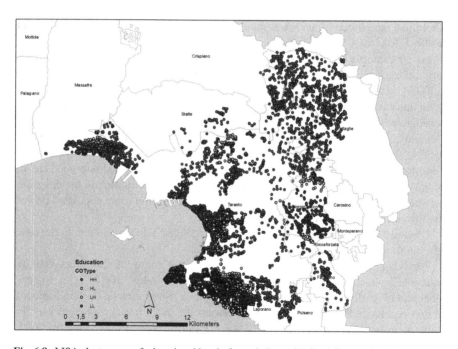

Fig. 6.8 LISA cluster map of educational level of population, with fixed distance band of 1,500 m

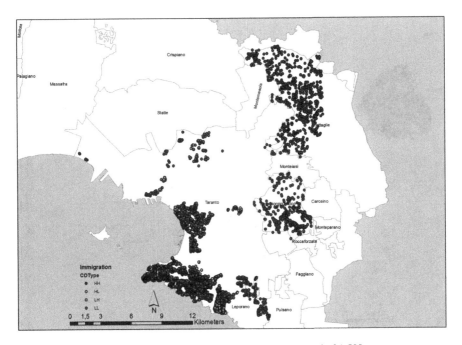

Fig. 6.9 LISA cluster map of immigration, with fixed distance band of 1,500 m

Analyzing the educational level of population (Fig. 6.8), the large clustering of "people which had never been to school or dropped out school without successfully completing primary school program" is mainly concentrated in the old part of the city with a high rate of elderly population and in the Talsano, Salinella, Tamburi and Paolo VI neighbourhood and in several municipalities surrounding Taranto (Montemesola, Grottaglie, San Giorgio Ionico, Faggiano) (Fig. 6.9).

Analyzing the cluster map of immigration the large clustering is mainly concentrated in the hold part of the city, with some hot spots in the same quarters of the city previously cited and in several municipalities surrounding Taranto (Grottaglie, San Giorgio Ionico, Leporano).

Unemployed clustering is more concentrated within the city of Taranto (Fig. 6.10).

6.4 Conclusions

This first essay to employ geo-statistical methods to identify spatial concentration of urban shrinking in micro census zones inside a medium city such as Taranto and in the nearest municipalities probably affected by phenomena of suburbanization has shown that it can be helpful to highlight quarters and directions of suburbanisation.

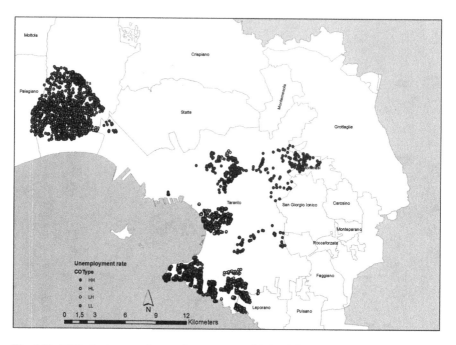

Fig. 6.10 LISA cluster map of unemployment rate, with fixed distance band of 1,500 m

It seems interesting the possibility offered by these analyses to deep the shrinking investigation inside the city, bringing out the quarters, the parts characterized by urban decline phenomena. In spite of having been used only six indicators, then a reduced number of indexes, in the case of Taranto, the study has indicated a probable presence of shrinking phenomena in the quarters which in other studies have been resulted to regenerate [26–28]. Tamburi, Paolo VI and Salinella quarters, in fact, together with old city center have all been the objective of regeneration policies.[7] In these quarters the economic situation is really difficult and the population decline is accompanied by a physical degradation, as already demonstrated in other studies [28]. It has also shown a direction for the suburbanization phenomena highlighting the census zone interested by the correlation.

The essential nature of the suburbs is that they are not constituted territories but dynamic, developing spaces undergoing slow, or even rapid transformation. Moreover, the suburbs are territories that cannot be completely remodelled: they can only be altered bit-by-bit, according to the degree of obsolescence of their different parts (uncultivated land, residual zones that can be bought up with public funds, partial

---

[7] The Salinella neighbourhood contract (€3 m), the Urban II program (€39 m), the Tamburi program agreement (€68 m), the Paolo VI program (€4 m), the Talsano program (€4 m) and Inner City Interventions (€6 m) are all autonomous programs with different urban objectives which are potentially useful [24].

urban renewal operations), or else according to political opportunity (planned transformation of a sector using public resources).

Geo-statistical methods could be effective to represent this suburbs dynamics, just analyzing data at a micro level. These tools could be more and more important for the larger cities where shrinking phenomena could be relevant inside the same city or the suburbanization could move inhabitants to the nearest municipalities, so increasing car mobility and in some case exalting shrinking phenomena.

References

1. Anselin, L.: GeoDa 0.9 User's Guide, p. 125. Spatial Analysis Laboratory, Department of Agricultural and Consumer Economics and CSISS, Urbana-Champaign Urbana (2003) http://geodacenter.org/downloads/pdfs/geoda093.pdf
2. Anselin, L.: Local indicators of spatial association – LISA. Geogr. Anal. **27**, 93–115 (1995)
3. Anselin, L.: Spatial Econometrics: Methods and Models. Kluwer, Boston (1988)
4. Audirac, I., Alejandre, J.A. (eds.): Shrinking Cities South/North. Juan Pablos Editor, México City (2010)
5. Badaloni, M., Vinci, E.: Contributi all'Analisi dell'Autocorrelazione Spaziale. Metron **46** pp. 119–140 (1988)
6. Bairoch, P., Batou, J., Chèvre, P.: La population des villes européennes de 800 à 1850. Droz, Centre d'histoire économique internationale, Genève (1988)
7. Beauregard, R.: Shrinking cities in the United States in historical perspective: a research note. In: Pallagst, K., et al. (eds.) The Future of Shrinking Cities – Problems, Patterns and Strategies of Urban Transformation in a Global Context, Monograph Series, pp. 61–68. Center for Global Metropolitan Studies, UC Berkeley (2009)
8. Bretagnolle, A., Mathian, H., Pumain, D., Rozenblat, C.: Long-term dynamics of European towns and cities: towards a spatial model of urban growth. Cybergeo: Eur. J. Geogr. Dossiers, 11ème Colloque Européen de Géographie Théorique et Quantitative, Durham, Royaume-Uni, 3–7 September 1999, Article 131. http://cybergeo.revues.org/566. Accessed 02 Jun 2012. doi:10.4000/cybergeo.566
9. Cheshire, P., Carbonaro, G.: Urban economic growth in Europe: testing theory and policy. Urban Stud. **33**, 1111–1128 (1996)
10. Coleman, D.: Immigration and ethnic change in low-fertility countries: a third demographic transition. Popul. Dev. Rev. **32**(3), 401–446 (2006)
11. Couch, C., Karecha, J., Nuissl, H., Rink, D.: Decline and sprawl: an evolving type of urban development – observed in Liverpool and Leipzig. Eur. Plann. Stud. **13**, 117–136 (2005)
12. Cunningham-Sabot, E., Fol, S.: Shrinking cities in France and Great Britain: a silent process? In: Pallagst, K., et al. (eds.) The Future of Shrinking Cities – Problems, Patterns and Strategies of Urban Transformation in a Global Context, Monograph Series, pp. 17–28. Center for Global Metropolitan Studies, UC Berkeley (2009)
13. Eberstadt, N., Groth, H.: Europe's Coming Demographic Challenge. Unlocking the Value of Health. American Enterprise Institute Press, Washington (2007)
14. Gilman, T.J.: No Miracles Here. Fighting Urban Decline in Japan and the United States. State University of New York Press, Albany (2001)
15. Grossmann, K., Haase, A., Rink, D., Steinführer, A.: Urban shrinkage in East Central Europe? Benefits and limits of a cross-national transfer of research approaches. In: Novak, M., Nowosielski, M. (eds.) Declining Cities/Developing Cities: Polish and German Perspectives, pp. 77–99. Instytut Zachodni, Poznan (2008)
16. Moran, P.A.P.: The interpretation of statistical map. J. R. Stat. Soc. B **10**, 243–251 (1948)

17. Murgante, B., Borruso, G.: Analyzing migration phenomena with spatial autocorrelation techniques. Lect. Notes Comput. Sci. **7334**, 670–685 (2012)
18. Murgante, B., Danese, M.: Urban versus rural: the decrease of agricultural areas and the development of urban zones analyzed with spatial statistics. Int. J. Agric. Environ. Inf. Syst. **2**(2), 16–28 (2011). doi:10.4018/jaeis.2011070102 (IGI Global)
19. Murgante, B., Las Casas, G., Danese, M.: Analyzing neigh bourhoods suitable for urban renewal programs with autocorrelation techniques. In: Burian, J. (ed.) Advances in Spatial Planning 165-178. InTech (2012). Open Access. ISBN: 978-953-51-0377-6, doi:10.5772/33747 http://www.intechopen.com/books/advances-in-spatial-planning/analyzing-neighbour-hoodssuitable-for-urban-renewal-programs-with-autocorrelation-techniques
20. O'Sullivan, D., Unwin, D.J.: Geographic Information Analysis. Wiley, Chichester (2003)
21. Oswalt, P., Rieniets, T. (eds.): Atlas of Shrinking Cities. Hatje Cantz Verlag, Ostfildern (2006)
22. Oswalt, P.: Shrinking Cities. International Research, vol. 1. Hatje Kanz Verlag, Hostfildern Ruit (2005)
23. Pallagst, K., Aber, J.: Introduction. In: Pallagst, K., et al. (eds.) The Future of Shrinking Cities – Problems, Patterns and Strategies of Urban Transformation in a Global Context, Monograph Series, pp. 17–28. Center for Global Metropolitan Studies, UC Berkeley (2009)
24. Perrone, N.: Vivere con la fabbrica. http://nodiossina.regione.puglia.it (2009). Accessed 7 Jan 2011
25. Rotondo, F., Selicato, F., Camarda, D.: Strategies for dealing with urban shrinkage: issues and scenarios in Taranto. Paper Presented to European Planning Studies in January 2012 (2012)
26. Rotondo, F., Perchinunno, P., Torre, M.C.: Estimates of housing costs and housing difficulties: an application on Italian metropolitan areas. In: Kis, S., Balogh, I. (eds.) Housing, Housing Costs and Mortgages: Trends, Impact and Prediction, pp. 93–108. Nova, New York (2010)
27. Rotondo, F., Selicato, F., Camarda, D.: Policies and strategies for dealing with different forms of shrinkage: the case of Taranto. In: Martinez-Fernandez, C., Kubo, N., Noya, A., Weiman, T. (eds.) Demographic Change and Local Development: Shrinkage, Regeneration and Social Dynamics, pp. 114–121. OECD Working papers series, Paris (2012)
28. Rotondo, F., Selicato, F.: The role of urban design in the process of regeneration of the suburbs: the case of the Puglia region. In: Panagopoulos, T., Noronha, T., Beltrao, J. (eds.) Advances in Urban Rehabilitation and Sustainability. World Scientific and Engineering Academy and Society (WSEAS) Pyreus, Greece (2010)
29. Scardaccione, G., Scorza, F., Casas, G.L., Murgante, B.: Spatial autocorrelation analysis for the evaluation of migration flows: the Italian case. LNCS **6016**, 62–76 (2010). doi:10.1007/978-3-642-12156-2_5
30. Oswalt, P.: Shrinking Cities, Vol.2, Interventions, Hatje Kanz Verlag, Hostfildern 663 Ruit (2006)
31. Tobler, W.R.: A computer movie simulating urban growth in the Detroit region. Econ. Geogr. **46**(2), 234–240 (1970)
32. Turok, I., Mykhnenko, V.: The trajectories of European Cities, 1960–2005. Cities **24**(3), 165–182 (2007)
33. Weber, L.: Demographic Change and Economic Growth. Physica-Verlag HD, Berlin (2010)
34. Wiechmann, T.: Conversion strategies under uncertainty in post-socialist shrinking cities – the example of Dresden in Eastern Germany. In: Pallagst, K., et al. (eds.) The Future of Shrinking Cities – Problems, Patterns and Strategies of Urban Transformation in a Global Context, Monograph Series, pp. 5–16. Center for Global Metropolitan Studies, UC Berkeley (2009)
35. Wu, C.-T., Zhang, X.-L., Cui, G.-H., Cui, S.-P.: Shrinkage and expansion in peri-urban China. Exploratory case study from Jiangsu Province. Paper Presented at the ACSP-AESOP 4th Joint Congress, University of Illinois at Chicago, 6–11 July 2008

Chapter 7
Social Identity as Determinant of Real Estate Economy in Manhattan

Carmelo M. Torre and Palmarita Oliva

Abstract This paper tells about a procedure for investigating the coherence of the relationship between a "wide" concept of spatial distance and the geographical variation of real estate value.

Such coherence is analyzed taking a special attention to the "multiple identity". That characterizes some urban places.

Many authors consider that real estate value of similar housing units can depend mainly on distance from some reference points; furthermore, its variation can be considered roughly linear.

On this basis the use of geo-statistical approaches based on kriging techniques has been developed in mass appraisal.

A second relevant point of view underlines the relationship between the presence of higher real estate value in those places where several amenities are coexisting.

But in those urban realities where the number of central points and the number of amenities are high, the complexity does not support the construction of models, and this complexity leads to a different concept of identity as synthesis of distance, borders and concentration.

In this complexity maybe further aspect can arise. In the case of study, that is to say New York, it is possible to investigate the effect of racial steering on ethnic dissemination and real estate variation. Born as a symbol of racial discrimination, in the nowadays city it assumes an identity character that affect in a singular way the

C.M. Torre (✉)
Department of Civil Engineering and Architecture, Polytechnic of Bari, Via Orabona 4, 70125 Bari, Italy
e-mail: torre@poliba.it

P. Oliva
EcoLogica, Corso Alcide De Gasperi 258, 70125 Bari, Italy
e-mail: poliva@eco-logicasrl.it

S. Montrone and P. Perchinunno (eds.), *Statistical Methods for Spatial Planning and Monitoring*, Contributions to Statistics, DOI 10.1007/978-88-470-2751-0_7, © Springer-Verlag Italia 2013

housing market. The aim of this paper is to demonstrate this hypothesis. [The paper is the result of a joint effort: Oliva wrote Sect. 7.1 and Torre wrote Sect. 7.2. The conclusions are due to both authors.]

Keywords Fuzzy clustering • Racial steering • Real estate market • Urban identity

7.1 Introduction

Real estate appraisal has founded its main approach on multiple regression analysis for a long time. Any kind of parameter has been investigated in the main urban reality of the world.

Furthermore, property value represents a major indicator of quality of life and services. In the 1970s, the concept of hedonic pricing has been pointed to define the relationship between the presence of the so-called *Amenities* (environment, urban services, cultural heritage) and the level of housing estate prices [4].

In the recent years, anyway, a new view of the relationship distance–estate value, put the attention on kriging techniques to make real estate value varying by the distance. Therefore, we can doubtlessly think that it is possible, and it is allowed by scientific literature the search for a model based on distance among settlements referring to some centrality.

But some limits of such models are identifiable with the aspects that will be reported as it follows.

Distance in urban complex realities can be measured towards/from a number of reference points, all potentially affecting real estate value.

A strong limitation regards the co-presence of many central amenities that makes difficult to identify the contribution of each one of the same amenities to the variation of value. In simple words the social complexity affects the value with a non-linear rule.

Last, but absolutely not the least, social identity of places affects real estate maybe more than physical distance.

In some places the urban social identity changes from road to road, the aspect of a quarter is totally different if compared with the neighbour.

In this approach, counterposed to purely statistical methods, relations of contiguity are investigated when it occurs inside a city that high residential areas and distressed areas coexist in proximity; the board among such two pieces of city is sometimes a physical "transition" element (a bridge, a road) by which you can move fast from one area to another without interruption.

"Public works" at the same time contribute to the construction of the grey area between quality and degradation, and those images people have of it once and for all and essentially due to its physical configuration, but the outcome is of the life stories of those who practice and the lives of their constant building and rebuilding perimeters and assignments to places.

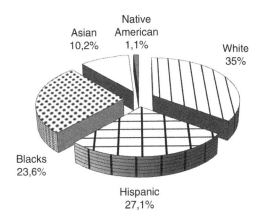

Fig. 7.1 Racial distribution in New York (modified from NY Bureau of Census, 2008)

Under the light of spatial diversity that is unlinked with distance, we can consider the evidence of New York City; its character is represented by the conjunction between the old metropolis of the 1930s, where the coexistence of Harlem with Chinatown and the cross road between the Fifth Avenue and the Broadway were so well described by Lewis Mumford [11] in his newspapers articles, and the new city where new ethnic groups are added to the old ones.

New York is a city of more or less 8,400,000 inhabitants. It covers an area of 1,214 km^2 at the mouth of the Hudson River in the Atlantic Ocean.

Situated partly on land and partly on islands in the Bay of New York (New York Bay) and is administratively divided into five districts (boroughs): Manhattan, Bronx, Queens, Brooklyn and Staten Island.

Of these, one is on the mainland (the Bronx, just north of Manhattan), three are located on an island surrounded by the sea (Staten Island, Queens and Brooklyn, respectively, in the northwestern and southwestern coast of the island of Long Island) and a Manhattan on appendix bottom of the peninsula where there is also the Bronx, but it is separated by Harlem River, river-canal linking the Hudson to the East River. The five borough offices are also metropolitan county: the county of New York itself occupies the whole of Manhattan, Brooklyn and Kings, that of Richmond Staten Island, the other two counties (Bronx and Queens) are homonyms of the boroughs whose administrative territory overlap.

The population of New York is among the most diverse in the world, both in terms of cultural ethnicity. Always a popular destination for immigrants from all over the world, today 36 % of the inhabitants were born abroad. Immigration recently seen at the top of the following countries: Dominican Republic, China, Jamaica, Guyana, Mexico, Ecuador, Haiti, Trinidad and Tobago, Colombia and Russia in the city there are about 170 different languages spoken. It also has the largest African American community in the USA (31 %), the largest Jewish community outside Israel (12 %) and the largest Puerto Rican community outside Puerto Rico.

The population is divided as shown in Fig. 7.1.

White, Hispanic, and Asian people represent about the 90 % of the population in New York. Only few native Americans and the 10 % of Asian complete the distribution of ethnic groups.

As regards the origin of white, the 8.7 % has Italian provenance, 6.9 % has Caribbean roots, 5.3 % Irish roots, 3.2 % Germany roots and 3.0 % Russian roots.

New York is the most populous city in the Union, and one of the most influential economic and cultural centres of the American continent in the global economy. New York is part of the system of "global economic network" whose development is influenced more by what happens in world events than by local [5]. As we can see below the city is characterized by relevant social problems due to the coexistence in one big area of a patchwork of ethnic groups that live in close contact and the coexistence of areas with different levels of quality resulting in urban and social issues that arise. The proximity of these areas and the accessibility of data has sparked interest in New York as a case study. The presence of physical boundaries such as bridges, roads and rivers that separate the different districts favouring the interest in comparing the phenomenon of urban poverty among the different districts that make up the City of New York.

The less fascinating, but considerably relevant aspect is that the variability of the urban context in New York is surely not gradually changing.

A sudden difference can arise between one avenue to another, in the high density, in the dynamic of buildings' substitution, decay, restoring, in a high density tissue; some buildings, some public spaces denote the immediate boundary between different neighbours that have a trade-marked identity.

The city is the symbol of metropolitan areas that grow both horizontally and vertically. The idea that the distance from a central position can be a measure of the value of a place at this point loses consistency. For example, such distance shall be measured horizontally and vertically. In cities where the urban grid in a vertical block contains more apartments than you can count from one side to another of a bridge suggests that the values of things do not vary gradually with distance from a service, from a park, from a central point; those values suddenly change, in a jump.

In addition, this jump of a knitted fabric to the other can pass from a block of a thousand of flats to a block as many offices, or by a block of five thousand white residents to five thousand Asian residents, and finally by a block of one thousand accommodations for rich to one thousand of popular housing.

The assumptions described above are intended to specify some special characters of the relationship between space and economic value of urban estate.

If we refer to the models of above (based on spatial density of amenities and distance), the high density, in the same time, denotes high concentration of possible amenities affecting real estate value; and the existence of physical boundary put on evidence a criticism for the idea of a gradual variation of values.

In other works we have already considered some criticism related to the assumption that given a unique segment of market, *ceteris paribus*, the only variable affecting the real estate value is the position [9, 10].

In this paper we assume that it is possible to define with a fuzzy clustering a similarity function in order to compare the districts inside the quarters of New York and to validate the relationship value–distance.

7.2 Measuring Values of Housing Property in New York

7.2.1 General Data of the Case of Study

Despite the physical continuity, the character of the city changes fast, especially when pedestrians pass from south Manhattan to Harlem along the Central Park, or from the Greenwich Village to the East side.

The city is facing with the greatest estate market crisis of the last two decades. Property prices are now at the same level of 15 years ago [3]. The refurbishment process shows some stop going around, and it is possible to discover some abandoned Building also near the seat of Wall Street.

All the city is divided in more less sixties administrative contexts, named Districts.

Manhattan accounts 12 districts, Brooklyn 18, Queens 13 and Bronx 12. In addition to the previous, Staten Island is divided into three districts that could be defined metropolitan only with some approximation.

The presence of public intervention is more spread than anybody could imagine.

In order to obtain a comparison, we developed several multidimensional analyses: starting from a measure of distress of housing stock and the related social policies, and from a measure of the ethnic distribution inside the City Quarters, we produced a study that shows that a main expressive social character of identity is identifiable in the multiracial dimension of New York.

The spatial transition from a racial concentration to another inside the urban tissue is almost instantaneous, due to the divisor character of some main streets in the city.

Aspects from which to define the level of urban social and housing distress are as follows:

- Family composition: couples vs wide families.
- High/low rent housing stock (under 50,000 dollars per year vs over 1,000,000).
- Mix of function inside the building (commercial, residential, offices).
- Real estate value.
- Racial composition and racial steering.

Racial steering is the attitude that addresses home buyers according to their race towards certain neighbourhoods. Racial steering practice can be identified in the action of advising customers to purchase homes in particular neighbourhoods on the basis of race.

The method will be to compare a couple of alternatives each other, which in this case are represented by the quarters' districts.

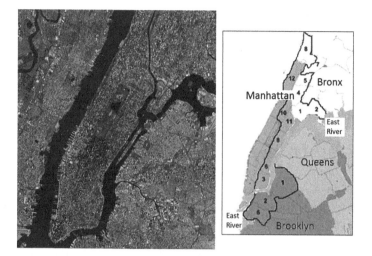

Fig. 7.2 Quarters surrounding Manhattan and the East River

7.2.2   Area of Study

The study interested the main boroughs of New York that profile the border between the main districts that make up the boroughs of Manhattan, Bronx Brooklyn and Queens. Substantially all those districts that are contiguous of Bronx Brooklyn and Queens to the East River, and those that belong to Manhattan.

The analysis does not consider data relating to the district of Staten Island.

The Quarters (Bronx, Queens, Brooklyn and Manhattan) with a high degree of reciprocal contiguity are physically filtered by the presence of physical constraints (bridges, canals) that mark the transition from one to another.

You can move from Manhattan to Queens in few minutes through the underground tunnel or from Brooklyn to Manhattan and back across the bridges, as many roads to allow structural grandeur of the passage between Manhattan and the Bronx continually doing their filter/connection function between the two parties.

It is not uncommon therefore to have socio-economic housing similarity or completely different in the two parts (Table 7.2).

The investigation has been carried out looking at a restricted group of districts that surround the core area of Manhattan. All the Eastern part of Manhattan was considered, from the southern to the northern area of the neighbour (Mn3, Mn6, Mn8, Mn10, Mn11, Mn12).

In addition, those districts that represent the enclosure of the Manhattan area have been considered; in the north side District of the Bronx; in the south area District of Brooklyn.

The Bronx's Districts (named Bx1, Bx2, Bx4, Bx5, Bx8) represent a physical area of continuity with Columbia and Harlem.

The Brooklyn's Districts (named Bx1, Bx2, Bx4, Bx5, Bx8) are joining the Skyscrapers' Peninsula by bridges.

Table 7.1 Distribution of rent of the housing property in the borderline districts

	Bronx 1+2	Bronx 4	Bronx 5	Bronx 8	Brook 1	Brook 2	Brook 6	Manh 3	Manh 6	Manh 8	Manh 10	Manh 11	Manh 12
Owner-occupied units	3,204	2,386	1,682	13,199	9,900	16,151	17,550	9,056	26,075	43,189	6,270	3,257	5,956
Less than $50,000	164	533	144	1,301	416	387	46	675	164	333	433	338	245
$50,000–99,999	66	402	84	499	202	402	50	229	66	270	148	44	113
$100,000–149,999	302	241	0	872	183	140	73	125	164	43	188	126	229
$150,000–199,999	191	169	29	1,323	338	173	111	102	267	257	52	19	191
$200,000–299,999	552	205	115	2,298	287	1,451	555	665	910	1,212	444	276	880
$300,000–499,999	1,478	569	931	329	1,485	3,362	1,999	174	5,697	5,455	854	736	2,498
$500,000–999,999	420	267	379	2,558	5,598	5,583	658	3,581	1,087	13,387	2,038	903	1,563
$1,000,000 or more	31	0	0	1,058	1,391	4,653	8,136	1,939	7,937	22,232	2,113	815	237

Source: NY Bureau of Census (2008)

Table 7.2 Distribution of value of the housing property in the borderline districts

	Bronx 1+2	Bronx 4	Bronx 5	Bronx 8	Brook 1	Brook 2	Brook 6	Manh 3	Manh 6	Manh 8	Manh 10	Manh 11	Manh 12
Occupied units paying rent	40,023	42,691	38,854	29,573	41,423	31,335	29,299	61,558	54,831	72,652	43,252	40,235	62,494
Less than $200	3,083	827	1,088	331	953	1,019	541	2,084	393	538	206	261	1,059
$200–299	5,661	2,958	248	707	3,047	2,325	1,263	6,026	669	1,022	4,333	6,149	2,517
$300–499	6,604	3,114	2,955	121	4,275	2,384	1,368	7,532	1,181	1,225	5,799	6,366	4,204
$500–749	81	7,185	5,798	3,724	6,831	4,706	2,842	13,684	2,416	4,356	10,756	9,314	10,462
$750–999	8,297	1,512	13,879	8,047	6,529	441	3,371	8,687	3,624	506	8,827	5,541	18,583
$1,000–1,499	6,701	12,067	10,923	11,902	10,619	6,217	6,459	9,475	13,348	14,822	6,912	574	19,592
$1,500 or more	1,577	142	1,731	3,652	9,169	10,274	13,455	1,407	332	45,629	4,565	4,515	6,077

Source: NY Bureau of Census (2008)

Table 7.3 Distribution of the housing property in the borderline-districts FY 2008 by mortgaged and rented stock

	Housing units with a mortgage (%)	Housing units without mortgage (%)	Occupied units paid rents (%)
Bronx 1+2	47.3	17.7	45.9
Bronx 4	40.9	14.6	51.8
Bronx 5	44.3	11.7	53.3
Bronx 8	22.5	15.2	37.1
Brook 1	5.6	24.6	40.5
Brook 2	29.6	14.8	30.6
Brook 6	24.0	15.9	33.0
Manh 3	36.2	16.4	37.9
Manh 6	23.7	13.6	30.3
Manh 8	25.4	13.9	29.5
Manh 10	36.3	19.0	39.7
Manh 11	30.4	19.9	30.9
Manh 12	19.6	8.6	44.5

All the area of study is posed along a south–north axis, for around 7 km.

Tables 7.1 and 7.2 show the main data regarding price values and rent values of districts in the case study.

Note that some districts of Brooklyn show median value of prices not dissimilar from the price of Manhattan area.

The analysis examines the relationships between the districts belonging to the quarter of Manhattan and districts belonging to the three quarters bordering it (Bronx, Queens, Brooklyn). Inside Manhattan, as we saw in the previous analysis, districts have a low aggregate poverty index.

More detailed analysis shows that Manhattan can be defined as a district is uneven from a social point of view with clear pockets of poverty at the neighbourhood such as Harlem and Chinatown that are opposed to areas more economically fortunate, such as Upper East Side which together help to reduce the overall poverty rate in the districts of Manhattan.

Secondly, Bronx is characterized by the widest presence of social housing, counterposed to a small part of properties in the free market: private properties are rented or are building units with a mortgage with a margin of more than 35 % of household income and therefore not easily accessible.

We can explore what real estate market indicates to us, and if there is a relationship between values housing typologies and property, and ethnic group.

By comparing the values of Brooklyn 2 and Queens 2 (Table 7.1), we can see that they are almost comparable; one might think that there is a similarity between the conditions of two seats belonging to adjacent districts.

A comparison among the values of Manhattan 10, 11 and 12 Bronx 1+2, 4 and 7 shows that the incidence of costs of living in case of mortgage or rent is higher in the Bronx than in Manhattan.

In the tenth district of Manhattan, the impact of the cost of living, in the presence of mortgage, and families in rent is just as high, and this is also in that piece of Manhattan district in between with Harlem, where the low income/black population is concentrated and construction stock is predominantly public.

Table 7.3 above shows the values of the impact of living, measured in relation to family income, and therefore the type of dwelling defined in real estate parlance, "non-affordable housing" and thus relating to the free market. Close examination of the data shows that it is possible to draw some considerations on the housing stock that characterizes the city of New York.

Manhattan is characterized by the widest presence of the occupied properties in the open market of rents.

7.2.3 Multidimensional Ranking

If the variation is so much connected to a high density of aspects in small areas, the dependence on these aspects is complex.

This complexity cannot be so easily broken down in simple relationships. At this point it is perhaps useful to consider the variables listed above as the size of a multi-variable fuzzy function.

For each alternative is composed of a set of criteria, representative of those given aspects of racial composition and housing stock.

Let's consider a fuzzy function μ that measures the dominance of an element X on an element Y and let's assume that:

- $\mu_\gg$ represents the fuzzy cluster of gaps that identify the highest level of prevalence in the comparison between X and Y.
- $\mu_>$ represents the fuzzy cluster of gaps that identify a moderate level of prevalence in the comparison between X and Y.
- $\mu_\sim$ represents the fuzzy cluster of gaps that identify a moderate level of similarity in the comparison between X and Y.
- $\mu_=$ represents the fuzzy cluster of gaps that identify the highest level of similarity in the comparison between X and Y.
- $\mu_<$ represents the fuzzy cluster of gaps that identify a moderate level of non-prevalence in the comparison between X and Y.
- $\mu_\ll$ represents the fuzzy cluster of gaps that identify the highest level of non-prevalence in the comparison between X and Y.

Let's consider a crisp function C that measures the dominance of an element X on an element Y and let's assume that:

- $C_\gg$ represents the crisp cluster of gaps that identify the highest level of prevalence in the comparison between X and Y.
- $C_>$ represents the crisp cluster of gaps that identify a moderate level of prevalence in the comparison.
- $C_\sim$ represents the crisp cluster of gaps that identify a moderate level of similarity in the comparison.
- $C_=$ represents the crisp cluster of gaps that identify the highest level of similarity in the comparison.
- $C_<$ represents the crisp cluster of gaps that identify a moderate level of non-prevalence in the comparison.

- $C_{\ll}$ represents the crisp cluster of gaps that identify the highest level of non-prevalence in the comparison.

The transition from moderate to highest is due to a threshold that is fuzzy for μ and crisp for C.

Through the "Novel Approach for Imprecise Assessment of Decision Environment" [2], which allows a fuzzy comparison based on multiple criteria comparison, we will evaluate the level of urban housing distress for each quarter.

From the evaluation, the approach produces a double ranking.

The first ranking expresses the values of $\Phi+$ corresponding to the possibility that a given district could be affected by urban distress.

The second rank, actually expresses the values of $\Phi-$, corresponding to the opposite possibility that a given district could not suffer for social distress of housing sock linked with racial composition.

Since fuzziness, $\Phi+$ and $\Phi-$ are not automatically reversible.

Under uncertainty it should be noted an asymmetry of values, and in consequence the two lists do not match.

Relations in formulas (7.1) and (7.2) show the expression of $\Phi+$ and $\Phi-$.

$$\Phi^+(X) = \frac{\sum_{k=1}^{n-1} [\mu_{\gg}(X, Y_k) \wedge C_{\gg}(X, Y_k) + \mu_{>}(X, Y_k) \wedge C_{>}(X, Y_k)]}{\sum_{k=1}^{n-1} C_{\gg}(X, Y_k) + \sum_{k=1}^{n-1} C_{>}(X, Y_k)} \tag{7.1}$$

$$\Phi^-(X) = \frac{\sum_{k=1}^{n-1} [\mu_{\ll}(X, Y_k) \wedge C_{\ll}(X, Y_k) + \mu_{<}(X, Y_k) \wedge C_{<}(X, Y_k)]}{\sum_{k=1}^{n-1} C_{\ll}(X, Y_k) + \sum_{k=1}^{n-1} C_{<}(X, Y_k)}. \tag{7.2}$$

If the observation of both lists expresses the same priority with respect to this alternative is also strengthened by the finding that the alternatives posed at the top (i.e. the "least worst" is the "best") are almost coincident, so there is a good level certainty widespread.

A breakdown of preferences confirmed by reference to different criteria, the priorities of the alternatives equally clear.

In our case the ranking $\Phi+$ and $\Phi-$ refer to the quality of settlement (typology and ratio between low price and high price properties). A simple correlation study in Table 7.4 allows to proceed with a further investigation.

The quality is associated with the mix of function (residential, commercial and services) to the housing typology (houses or dwelling in multi-family buildings).

$\Phi+$ and $\Phi-$ appear strictly connected with the distribution of median price value and in the same with the presence of Hispanic, White and Native American (the absolute value of the correlation index is mostly for all the groups equal or higher than 0.7). This correlation should allow us to think that there is some evidence of the relationship between racial composition of the population quarter and real estate value.

Table 7.4 Correlation between quality, values and racial group

	Median value	1–2 familiar house	Multi familiar house	Commerce-offices	[<50,000]/[>1 million]	$\Phi+$ housing stock	$\Phi-$ housing stock
Median value	1.00	−0.04	−0.37	0.33	−0.63	0.71	−0.71
1–2 familiar house	−0.04	1.00	0.74	−0.41	0.35	−0.25	0.26
Multifamiliar house	−0.37	0.74	1.00	−0.61	0.51	−0.72	0.73
Commerce-offices	0.33	−0.41	−0.61	1.00	−0.36	0.65	−0.66
[<50,000]/[>1 million]	−0.63	0.35	0.51	−0.36	1.00	−0.74	0.71
$\Phi+$ housing stock	0.71	−0.25	−0.72	0.65	−0.74	1.00	−0.96
$\Phi-$ housing stock	−0.71	0.26	0.73	−0.66	0.71	−0.96	1.00
White	0.07	−0.33	−0.49	0.76	−0.51	0.61	−0.69
Black	−0.29	0.69	0.55	−0.34	0.02	−0.16	0.20
Asian	0.25	0.38	−0.06	0.31	−0.19	−0.02	0.02
Other non-Hispanic	0.36	0.07	0.00	0.63	−0.27	0.22	−0.31
Hispanic	−0.69	0.25	0.45	−0.32	0.89	−0.78	0.72
Native American	0.56	−0.14	−0.53	0.42	−0.84	0.82	−0.86
Foreign born	−0.56	0.14	0.53	−0.42	0.84	−0.82	0.86

If we consider that the spatial transition from a racial prevalence to another is connected as well with quarters, this link appears more strong that the one linked to the distance. Maybe it could be still existing a sport of racial steering that concentrates some ethnic group, still homogeneous in terms of family income and general economic conditions and attitude to expenditure, in some area with some real estate value.

A further proof of this connection could be shown be the absence of a strong relationship of real estate value with the diffusion of black population.

Maybe black population nowadays, in the USA of Obama, is equally present in high, middle and low class.

7.2.4   Clustering

In the fuzzy cluster method [7, 12] the difference of value between elements of a cluster is measured on the basis of semantic distance [4].

Formula (7.3) shows the formulation of the semantic distance among districts, measured in a multidimensional space where on the axes are measured the defined criteria referring to housing characters and racial steering effects [3, 6]. The "Semantic Distance" is represented by the sum of two double integrals:

$$S_d(f_j(x), g_j(y)) = \int_{-\infty}^{+\infty} \int_X^{+\infty} |Y - X| g_j(Y), f_j(X) dY\, dX + \int_{-\infty}^{+\infty} \int_{-\infty}^{X} |X - Y| f_j(X), g_j(Y) dY\, dX.$$

(7.3)

Let's give a jth quantitative attribute of a set of two elements X and Y; let's suppose that $f_j(X)$ and $g_j(Y)$ represent the value functions of the fuzzy attribute for X and Y.

The functions f_j and g_j can be crisp numbers (this means that the function give a certain result), probabilistic values (this means that f_j and g_j represent expected values), or fuzzy numbers (this means that f_j and g_j represent ownership function).

The "Semantic Distance" provides an indication of the certainty of the prevalence of one alternative over another. It, between two points identifying two different levels of courts, may take a value, at the comparison between judgements of certain preference.

After the clustering a *Similarity index* is calculated. The index aggregates the group on the basis of two indicators: price value and rent value. Table 7.5 shows the matrix of similarity among all couples of districts. Furthermore, the dendrogram in Fig. 7.3 shows the relevant similarities.

In the case of study some relevant clusters have been identified as significant of toblerian condition of contiguity of values: the couple of the eleventh and tenth District of Manhattan represent a Grey area where Manhattan reality is melt with

Table 7.5 Similarity matrix among districts

	Bronx 1+2	Bronx 4	Bronx 5	Bronx 8	Brook 1	Brook 2	Brook 6	Manh 3	Manh 6	Manh 8	Manh 10+11	Manh 12
Bronx 1+2	1.000											
Bronx 4	0.697	1.000										
Bronx 5	0.812	0.699	1.000									
Bronx 8	0.773	0.697	0.859	1.000								
Brook 1	0.806	0.663	0.859	0.852	1.000							
Brook 2	0.769	0.661	0.850	0.868	0.868	1.000						
Brook 6	0.728	0.600	0.779	0.788	0.829	0.844	1.000					
Manh 3	0.806	0.663	0.859	0.852	0.880	0.871	0.830	1.000				
Manh 6	0.675	0.642	0.779	0.829	0.788	0.844	0.811	0.788	1.000			
Manh 8	0.644	0.573	0.718	0.745	0.755	0.794	0.854	0.755	0.831	1.000		
Manh 10+11	0.868	0.661	0.807	0.741	0.819	0.748	0.732	0.819	0.679	0.664	1.000	
Manh 12	0.812	0.699	0.870	0.862	0.862	0.855	0.780	0.862	0.780	0.717	0.808	1.000

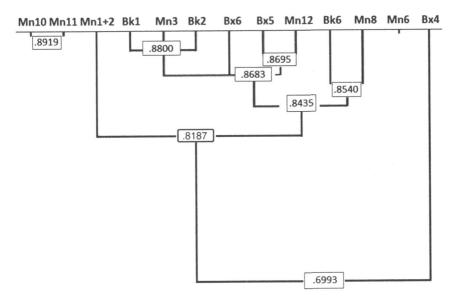

Fig. 7.3 Dendrogram of similarity for New York Districts

the Bronx in correspondence with the poorest part of Harlem (similarity equal to 0.89)

Another relevant group is represented by some district of Brooklyn connected with the corresponding district of Manhattan from one side to another of the east river. These realities represent the richest part of the "skyscrapers area".

Bk1 and Bk2 of Brooklyn and the Mn3 of Manhattan (similarity 0.88)

The richest part of Manhattan (District Mn8) is near by Guggenheim Museum, towards the east side, moving from the Fifth Avenue and Central Park.

Other group looks less spatially related. In front of such part we find the richest district of Brooklyn (District Bk6) that is shifted more towards south. The similarity decreases (0.85).

7.3   Conclusions

The use of fuzzy multidimensional analysis gives back an idea of the possible relationship between values and identity of quarters in the New York reality. The identity can be marked by ethnic distribution. The qualitative idea of contiguity, instead, in the definition of similarity appears more strongly than the quantitative idea of distance.

In fact the similarity identifies immediately some areas that are on the borderline of neighbours and that show a proper identity (as the case of the poorer Harlem, or the richer South–East Side).

More than among quarters, it seems more appropriate to consider at the local scale inside each district or inside each—however defined—part of the neighbours, the influence of distance.

The reality of the city shows a variation that is well identified with the physical/perceptual modern/historical barriers that still design the urban tissue.

Even if some parts of the city have changed their social character, and face a continuous dynamic substitution of some ethnic group with another, the structural and dimensional character of property seems to survive as evidence of a different economic status of residents in some area.

Such estate values, in their highest peaks, are often conserving a "status symbol" [1, 8], while in the intermediate conditions their variation maintains a character of fuzziness.

Finally the clustering appears useful to put on a better evidence the identicalness of some well-characterized areas, evidence to the walker, as Lewis Mumford told.

References

1. Aldrich, H., Cater, J., Jones, T., Mc Evoy, D., Velleman, P.: Ethnic residential concentration and the protected market hypothesis. Soc. Forces **63**(4), 996–1009 (1985)
2. Munda, G.: Social Multicriterial Evaluation for a Sustainable Economy. Springer, Berlin (2008)
3. Portes, A.: Social capital: its origins and applications in modern society. Annu. Rev. Sociol. **24**, 1–24 (1998)
4. Rosen, S.: Hedonic prices and implicit markets: product differentiation in pure competition. J. Polit. Econ. **82**(1), 34–55 (1974)
5. Sassen, S.: Cities and communities in the global economy. Am. Behav. Sci. **39**(5), 629–639 (1996)
6. Smith, C.: Asian New York: the geography and politics of diversity. Int. Migr. Rev. **29**(1), 59–84 (1995)
7. Takahashi, K., Tango, T.: A flexibly shaped spatial scan statistic for detecting clusters. Int. J. Health Geogr. **4**, 11–13 (2005)
8. Thomas, J.M.: Planning history and the black urban experience: linkages and contemporary implications. J. Plan. Educ. Res. **14**(1), 1–11 (1994)
9. Tobler, W.R.: A computer movie simulating urban growth in the Detroit region. Econ. Geogr. **46**(2), 234–240 (1970)
10. Torre, C.M., Mariano, C.: Analysis of fuzzyness in spatial variation of real estate market: some Italian case studies. In: Phillips-Wren, G., Jain, L.C., Nakamatsu, K., Howlett, R.J. (eds.) Advances in Intelligent Decision Technologies, pp. 269–277. Springer, Berlin (2010)
11. Wojtowicz, R. (ed.): Sidewalk Critic: Lewis Mumford's Writings on New York. Princeton Architectural Press, New York (2000)
12. Zadeh, L.A.: Fuzzy sets. Inf. Control **8**(3), 338–353 (1965)